Hydronics

Disclaimer

This book can only be considered a general guide. The author and publisher have neither liability nor can they be responsible to any person or entity for any misunderstanding, misuse or misapplication that would cause loss or damage of any kind, including material or personal injury, or alleged to be caused directly or indirectly by the information contained in this book.

HYDRONICS
The Art of Cooling and Heating With Water

George H. Clark

Edited by
Harry D. Busby

Business News Publishing Company
Troy, Michigan
USA

HYDRONICS
The Art of Cooling and Heating With Water

George H. Clark

Edited by
Harry D. Busby

Business News Publishing Company
Troy, Michigan

Library of Congress Cataloging in Publication Data

Clark, George H.
Hydronics: the art of cooling and heating with water.

Reprint. Originally published: Des Plaines, IL : Nickerson & Collins, 1973.
Includes index.
1. Air conditioning. 2. Hot-water heating.
3. Refrigeration and refrigerating machinery.

I. Busby, Harry D. II Title.
TH7687.C53 1986 697.9'3 86-4174
ISBN 0-912524-37-5

Printed in the United States of America

HYDRONICS

Fundamentals

The term "Hydronic Cooling" implies cooling with water. In effect water is used as a heat carrier to transport heat from an air cooling coil to a refrigerant evaporator. The water in a closed system can be readily pumped from one heat exchanger to another to pick up the heat from higher temperature air and to give up the heat to lower temperature refrigerant.

Water serves as an excellent heat carrier as it has the advantage of a high specific heat, low viscosity and is non-corrosive to most materials although it may serve as a catalyst to promote rusting of iron pipe when oxygen is present. To prevent corrosion in iron pipe, closed water systems must be used so that the free oxygen which may be present in water cannot be replaced to cause continuous rusting or corrosion. In order to have a closed chilled water system an expansion tank is necessary to maintain a reasonably constant pressure in the system as the volume of water increases or decreases with temperature changes.

For chilled water systems the volume of the water will be at a maximum at 100°, assuming it ever warms up to that temperature, and will be a minimum at about 40°F. The variation in volume from 40°F to 100°F is only about .7% and an expansion tank having a volume of 3.0% of the volume of the chilled water system is adequate. However, if the same system was to be used for hydronic heating a much larger expansion tank would be required. The variation in water volume from 100°F to 250°F is over 16% and an expansion tank volume of 22% of the total system volume is recommended.

Water used as a heat carrier in hydronic cooling does not cause bubble formation such as occurs when used in hydronic heating where bubbles are liberated as the water passes through a boiler. However, when water is used in hydronic cooling systems, air separating devices are required to prevent circulation of trapped air which would otherwise cause noise and possible pump damage. Since hydronic cooling and heating systems may be combined so as to use the same final heat exchanger surface any discussion of hydronic cooling must point out the similarities and differences in the summer and winter aspects of air conditioning.

Heating and Cooling Loads

To compare hydronic cooling with hydronic heating a small commercial building is used as an example. Assume a rectangular 100 x 300 building with wall 20 high and having a flat roof. Also assume that the window and door area is 10% of the wall area and that the infiltration of air into the building amounts to two volume changes per hour for both summer and winter. Assume also that there are 100 people working in the buildnig during the day and that electric lights average 3 watts per square foot of floor area, and that 20 hp. of electric motors are used during working hours. Heat transmission factors for walls are taken as .25 Btu per hr. per ° td. per sq. ft. and for the windows the "U" factor is 1.13. The roof "U" is taken as .20.

The great heating load occurs at night with no people in the building and no lights on. The entire heating load is made up of heat loss through the walls, windows, and roof and the fresh air load which is the amount of heat required to raise the temperature of the air, which leaks in at outside temperature, to the inside temperature. To get a condition where wall heat transmission is the same for heating and cooling a 20° difference between inside and outside is used to start with and the heating load would then be calculated as follows:

1. Heat transmission through walls = AU x °TD where:
   A = wall area in square feet
   U = wall heat transmission factor Btu per hr per °td per sq. ft.
   TD = inside temperature — outside temperature °F.

   Aw = net area of 4 walls — 10% window area = .90 x gross wall area = .9 2W + 2L) H where W = width of building, L = length, H = Height

   Aw = (2 x 100 + 2 x 300) H x .90 = 800 x 20 x .90 = 14,400 sq.ft.

   Ag = (glass area) = (2 x 100 + 2 x 300) H x .10 = 800 x 20 x .10 = 1600 sq.ft.

   Ar = (roof area) = W x L = 100' x 300' — 300,000 sq.ft.

Heat transmission loads at 20 °td =

(a) For walls

$$\frac{\overset{(A)}{14{,}400\ \text{sq.ft.}} \times \overset{(U)}{.25\ \text{Btu}} \times \overset{(TD)}{20\ ^\circ\text{td}}}{\text{sq.ft.} \times {}^\circ\text{td} \times \text{hr.}} = 72{,}000\ \text{Btu hr.}$$

(b) For glass

$$\frac{\overset{(A)}{1{,}600\ \text{sq.ft.}} \times \overset{(U)}{1.13\ \text{Btu}} \times \overset{(TD)}{20\ ^\circ\text{td}}}{\text{sq.ft.} \times {}^\circ\text{td} \times \text{hr.}} = 36{,}160\ \text{Btu hr.}$$

(c) For roof

$$\frac{\underset{(A)}{30{,}000 \text{ sq.}} \times \underset{(U)}{.20 \text{ Btu}} \times \underset{(TD)}{20\ ^\circ\text{td}}}{\text{sq.ft.} \times {}^\circ\text{td} \times \text{hr.}} = 12{,}000 \text{ Btu hr.}$$

Total heat transmission load at 20 °td = 228,160 Btu hr.

This heat transmission load would apply to the example building for maintaining 70° temperature inside the building continuously with 50° outside temperature and not considering any other loads. This load would be a normal cooling load with respect to transmitted heat if the inside temperature was 70° while the outside was 90° all night. The location of the building or the direction it faces does not enter into this transmission load.

However the largest cooling load occurs in the daytime when the sun is shining on the building. This additional heat transmission load is called the solar load and must take into account the location of the building (latitude) the direction the building faces (N.S.E. or W.) the time of year, the time of day, the color of the walls, and how the windows are shaded or if they have venetian blinds etc.

Fig. 1 shows the building assumed to be at 40° latitude (about that of Cincinnati) and the largest load is at midsummer. The building is taken as red brick and having medium light awnings over east, south and west windows and the roof is taken as tar and gravel. Tables 1, 2 and 2A will be used in determination of the solar load which is calculated as follows:

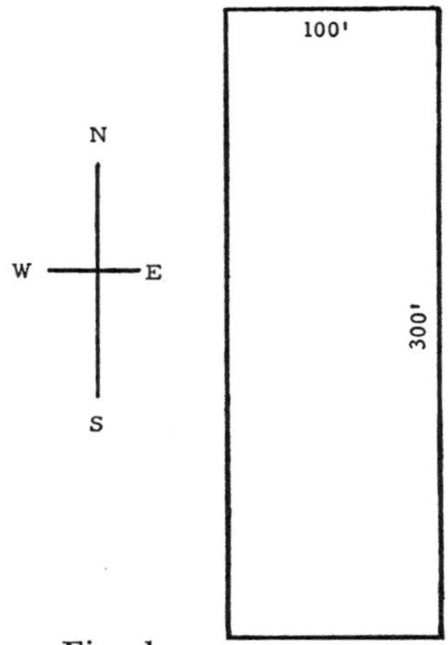

Fig. 1

Table 1 — Solar Heat Intensity at Various Latitudes and Orientations

30° LATITUDE							*35° LATITUDE*				
A.M.		*N.E.*	*E*	*S.*	*S*	*Horizontal*	*N.E.*	*E*	*S.E.*	*S*	*Horizontal*
read down	*P.M. read up*	*N.W.*	*W*	*S.W.*	*S*	*Horizontal*	*N.W.*	*W*	*S.W.*	*S*	*Horizontal*
5	7	—	—	—	—	—	9	9	3	—	—
6	6	47	51	24	—	9	67	72	35	—	15
7	5	136	160	90	—	68	142	174	103	—	77
8	4	151	205	136	—	147	150	209	145	—	151
9	3	127	189	140	8	214	118	191	154	26	214
10	2	79	141	122	31	265	60	143	139	55	265
11	1	21	78	85	45	296	2	75	103	72	291
12	12	—	—	36	50	305	—	—	55	78	300

40° LATITUDE							*45° LATITUDE*				
A.M.		*N.E.*	*E*	*S.*	*S*	*Horizontal*	*N.E.*	*E*	*S.E.*	*S*	*Horizontal*
read down	*P.M. read up*	*N.W.*	*W*	*S.W.*	*S*	*Horizontal*	*N.W.*	*W*	*S.W.*	*S*	*Horizontal*
5	7	14	14	5	—	—	25	24	9	—	—
6	6	72	80	40	—	19	89	99	52	—	26
7	5	143	180	112	—	82	149	194	125	—	90
8	4	143	211	155	8	152	140	219	171	22	156
9	3	104	192	168	46	213	92	194	183	65	210
10	2	46	143	156	77	258	33	144	171	98	251
11	1	—	75	121	95	284	—	75	139	121	274
12	12	—	—	73	103	293	—	—	91	128	282

The solar heat load through a wall or roof is taken as S.L. = A x Fe x Ft x I and the solar load for a window is taken as A x Fs x I where:

S.L. = the solar load for the area considered in Btu per hr.

A = the area under consideration in square feet.

Fe = the color factor or the decimal part of the solar heat which is absorbed by the outer surface of the wall or roof. The balance of the heat is reflected away. See Table 2.

Ft = a transmission factor or the decimal part of the solar heat absorbed by the outer wall which is transmitted through the wall. The balance is given to outside air, or radiated to other outside objects.

Table 2 — Solar Heat Absorption Coefficient "Fc"

Surface Color	*Finish or Material*	*Absorption Coefficient*
Aluminum Silver	Polished Mirror Finish	.1
White	Enamel, gloss paint, etc. Marble	.2
Light Colors	Flat Painted	.3
Light Color Off White	Flat Finish, Painted Wood	.4
Medium Light gray, green, yellow, etc.	Painted Flat	.5
Medium Light	Face Brick — Rough Surface	.6
Medium Dark Colors	Gravel Over Tar Roofs Medium Red Brick Medium Gray Stone Black Polished Marble Black Enamel	.7 .7
Dark Colors	Flat Paints	.8
Very Dark or Black	Flat Black Paint Dark Rough Surfaces	.9

I = the solar intensity on a surface from table 1. Btu per hr per sq.ft.

Fs = shading factor or decimal part of solar heat which enters through transparent areas depending on type and color of window shading. See table 2A.

Ft = depends upon the U factor of the wall and is estimated as 23% of the U factor for wall or roof.

All net wall areas are taken as 90% of the gross wall area and the glass is taken as 10% of the gross wall area. (For more refined calculations a time lag through the walls should be considered also but for sake of simplification is not considered here.)

Solar load calculations for example building at 3 pm (Sun Time)

1. South wall

   A Fc Ft I

   .90 x (100 x 20) x .7 x .23 x .25 x 46 = 3.340 Btu per hr.

2. South glass

   A Fs I

   .10 x (100 x 20) x .25 x 46 = 2,300 Btu per hr.

3. West wall

A Fc Ft I

.90 x (300 x 20) x .7 x .23 x .25 x 192 = 41,750 Btu per hr.

4. West glass

A Fs I

.10 (300 x 20) x .25 x 192 = 288,800 Btu per hr.

5. Roof

A Fc Ft I

100 x 300 x .7 .23 x .20 213 = 205,500 Btu per hr.

281,690 Btu per hr.

The comparison of building loads can now be made for the heat transmitted through walls, roof, and glass areas. In the summer load a 20° td. is about standard or 75° in 95° weather. However for heating to maintain an 80° td. (70° inside −10° outside) the heat transmitted due to difference between inside and outside air temperatures would be 4 times as great in the winter condition as it would be in the summer condition so the winter and summer building loads compare as:

(a) Summer (due to 20 °td)
228,160 + (solar) 281690 = 509,850 Btu per hr.

(b) Winter (due to 80 °td)
228,160 x 4 = 912,640 Btu per hr.

The only additional heating load is the air infiltration load while for cooling the additional loads are the air infiltration or fresh air load plus the internal load. The fresh air load is based on the assumption of 2 complete air changes per hour. The pounds of air contained in the building can be based on the weight of "standard air" which weighs

Table 2A — Window Shading Factor "Fs"

Type of Shading	*Reflecting Surface Color*	*% Solar Heat to Inside*
None	Polished Glass—Single Pane	90%
None	Polished Glass—Double Pane	85%
Awnings	Very Light	20%
Awnings	Medium Light	25%
Awnings	Dark	30%
Venetian Blinds	Bright Light Color or Aluminum	55%
Venetian Blinds	Light Flat Painted or Dark Polished	60%
Venetian Blinds	Dark Flat Painted	65%
Window Shades (drawn)	Aluminized—Bright	50%
Window Shades	Light Flat Shades	65%
Window Shades	Dark Flat Shades	80%

.075 pounds per cubic foot. The specific heat of air at constant pressure is taken as .241 Btu per pound per degree temperature change (t.c).

The volume of the building is 100 x 300 x 20 = 600,000 cu.ft.

The weight of the air in the building is 600,000 x .075 = 45,000 lb.

The amount of fresh air per hour is 45,000 x 2 changes per hour = 90,000 lb per hr.

The fresh air load then is: 90,000 lb per hr x 80 °tc x .241 Btu per lb per tc = 1,736,000 Btu per hr.

This is all sensible heat.

For the cooling season based on the same amount of air the sensible heat would only be 25% of that for heating because it would only have to be cooled 20° as compared to an 80° rise for heating. The sensible load for cooling then is .25 x 1,736,000 Btu per hr. = 434,000 Btu per hr.

However if the outside air is at 95° dry bulb and 75° wet bulb a psychrometric chart shows that it contains about 38.5 Btu per lb and if the inside air is to be maintained at 75° and 50% relative humidity the chart shows it has a wet bulb temperature of about 62.5° and contains about 28.2 Btu per lb and the total hourly fresh air load is 90,000 lb per hr. x (38.5 – 28.2 = 90,000 x 10.3 = 927,000 Btu per hr.

The latent heat load is the total fresh air load of 927,000 – 434,000 sensible load or 493,000 Btu per hr. latent load.

The total building and fresh air loads for heating 80° and for cooling 20° now compare as in Table 2B.

Table 2B	*Heating*	*Cooling*
1. Transmission due to difference between inside and outside temperatures	912,640	228,160
2. Solar Heat	0	281,690
3. Fresh air – sensible	1,736,000	434,000
4. Fresh air – latent	0	493,000
Total building and fresh air loads	2,648,640	1,436,850

The internal loads due to heat generating devices in the room add to the cooling load and would detract from the heating load except that they are not continuous. The heat input due to 100 people taken as equivalent to "Standing – Light Activity" Table 3 is

100 x 235 = 23,500 Btu per hr. of sensible heat and

100 x 265 = 26,500 Btu per hr. of latent heat for a total of 50,000 Btu per hr.

The heat input due to 20 hp of electric motors at full load and an estimated efficiency of 80% will be:

$$\frac{.80 \text{ ME x hp} - \text{HR}}{20 \text{ hp x } 2545 \text{ Btu}} = 63{,}600 \text{ Btu per hr, and is all sensible heat.}$$

The heat input due to electric lights based on 3 watts per square foot of floor area and 100 x 300 = 30,000 square feet is:

$$30{,}000 \text{ sq.ft.} \times 3 \text{ watts} \times \frac{3.413 \text{ Btu}}{\text{watt hr.}} = 307{,}170 \text{ Btu per hr.}$$

Adding these loads to the building and fresh air loads the total loads add up as in Table 3A.

Table 3 – Heat Rejected by People at Various Activity Rates

Degree of Activity	*Sensible Heat Btu/hr.*	*Latent Heat Btu/hr.*	*Grains of Moisture Evaporated Per Hour*
Dreamless Sleep	175	65	440
Restless Sleep	200	130	875
Sitting at Rest	220	180	1210
Sitting–Reading, Theater	225	200	1350
Sitting–Moderate Office Work	225	265	1790
Sitting–Active Office Work, Typing	235	280	1890
Standing at Rest	225	215	1450
Standing–Light Activity	235	265	1790
Standing–Active Clerk	250	350	2360
Slow Walking or Dancing	260	440	2960
Moderate Walking or Dancing	300	600	4040
Busy Waiters	325	675	4550
Fast Walking	375	750	5050
Trotting	500	1000	6740
Fast Running or Vigorous Exercise	700	1500	10,100
Maximum Effort (Short Period)	900	2100	14,140

NOTE: The ratio of latent to sensible heat rejection varies with the ambient air temperature and humidity. The above data is for estimating purpose. At ambient temperatures exceeding 100°F. essentially 100% of the heat rejected would be latent heat.

The total load for heating in this example is about 43% more based on 80° td. than for the cooling load based on 20° td. For home and small commercial heating systems the temperature reduction of the water through the heating coils may average about 20° although in large commercial and industrial buildings smaller piping may be used when 60° to 80° water temperature reduction is taken through the coils. For hydronic cooling systems the temperature rise through the coils and the temperature reduction through the chiller is usually taken as 8 to 10° td. so that the water flow rate must be greater for the same rate of heat transfer when cooling than it is for heating.

Table 3A

		Heating (80 °td)	Cooling (20 °td)
1.	Heat transmitted due to °td	912,640	228,160
2.	Solar heat		281,690
3.	Fresh air load—sensible	1,736,000	434,000
4.	Fresh air load—latent		493,000
5.	People—sensible		23,500
6.	People—latent		26,500
7.	Motors (20 hp)		63,600
8.	Electric lights		307,170
	Totals	2,648,640	1,857,630

In heating the entire load is a sensible heat load except where humidifiers are used in which case some additional load is added to vaporize water and pass it into the building.

Adding moisture to the air during the heating season in an attempt to keep indoor humidity up above 30 to 40% is not practical when cold outside temperatures prevail. At 75° indoor temperature and 30% relative humidity the dew point of the indoor air would be about 42°F. and with outdoor temperatures down to 0° a single pane window with high wind velocity would be much less than freezing temperatures and would remove the moisture from the building as it turns into frost on the windows and with a dewpoint as high as 42° moisture could condense on cold walls if it didn't have the colder windows to condense on.

In cooling, from 10% to over 50% of the total cooling load may be latent load. In the example building the latent load due to fresh air and people totalled 519,500 Btu per hour or approximately 28% of the total load. Most of the air conditioning systems in common use do not use any humidity control systems and depend upon the coil temperatures to be such as to provide approximately 50% to 55% relative humidity. This is usually possible with water temperatures entering the cooling coil at about 45° and rising to 55° in the coil. An increase in inlet water temperature of 5° decreases the amount of dehumidification that can be obtained in a coil considerably more than it decreases the sensible cooling that can be done and will raise the building relative humidity. This in turn requires maintaining lower dry bulb temperature to produce comfort conditions.

Since a ton of refrigeration is a heat transfer rate of 200 Btu per minute and a gallon of water weighs 8-1/3 pounds or 25 lb per 3 gallons it follows that for each ton of refrigeration produced by chilled water the product of the gpm. circulated times the °td. change in the coil (or chiller) is:

200 Btu per minute = gpm x 25 x °td and gpm x °td =

$$\frac{200}{25/3} = \frac{200 \times 3}{25} = 24$$

so that the relationship of flow rate and °td is as in Table 3B.

Table 3B

	Flow Rate	°T.D.	Flow Rate	°T.D.
2	gpm per ton	12°	4 gpm per ton	6°
2.4	gpm per ton	10°	6 gpm per ton	4°
3.0	gpm per ton	8°	8 gpm per ton	3°

For the building used as an example the total cooling load is

$$\frac{1{,}857{,}630 \text{ Btu per hr.}}{12{,}000 \text{ Btu per Ton-hr.}} = 155 \text{ tons}$$

The rate of water circulation can be determined as the heat transfer rate in tons multiplied by the flow rate in terms of gpm per ton. It may be determined from the following:

$$\text{Gpm} \times 500 \times {}^\circ\text{td} = \text{Btu per hr. or gpm} = \frac{\text{Btu per hr.}}{500 \times {}^\circ\text{td}}$$

based on the heating load and 20 °td water and the cooling load and 8 °td water the hydronic flow rates for heating and cooling would be:

(a) Heating gpm $= \dfrac{2{,}648{,}640 \text{ Btu per hr.}}{500 \times 20} = 265$ gpm

(b) Cooling gpm $= \dfrac{1{,}857{,}630 \text{ Btu per hr.}}{500 \times 8} = 464.4$ gpm

Where the same piping is to be used for both heating and cooling the cooling requirement, as in this case, is the one that determines the pipe and pump requirements for hydronic systems.

Piping Systems

Installation and service engineers who are going to perform their services on hydronic cooling systems should have a basic knowledge of hydronics so as to be able to analyze and effect cures for the troubles that may occur. The man well-versed in the art of refrigeration and controls should know that the proper operation of the water chiller depends to some extent on the original piping arrangement and the operation of the air chilling unit. Persons knowledgeable in hydronic heating must realize that the problems in hydronic cooling have similarities and distinct differences with respect to hydronic heating.

Essentially in hydronic heating the major function is to add sensible heat to the space being heated. In cooling, the removal of latent heat to remove excess moisture from the air may approach in importance, the removal of sensible heat to lower the air temperature so that a basic knowledge of psychrometry is helpful. In heating, water temperatures up to 250°F and higher, provide a high temperature differential to heat a space up to 70 or 75°. In cooling, water at 40 to 50° is only a comparatively few degrees below the 75° space temperature which may be required. It is only a very few degrees below the dew point of 50° which may be required in the conditioned space.

In hydronic heating, the temperature of the water as the heat carrier can be very high. However, as a heat carrier in hydronic cooling it is impractical to work with water below 40°F (10° below desirable dew points) due to the chance of freezing in the chiller. If an additive is used to lower the freezing temperature of the brine, formed as a result of the anti-freeze material added, there is still the limitation which can be caused by coil icing which blocks off air flow through the coils.

Properly designed hydronic heating systems are closed systems using surface type heat exchangers so that the water does not come in contact with air (except in the expansion tank) so that there is no continuous supply of oxygen to rust iron or iron alloys in the piping, pumps and boilers. In most instances this is also true of hydronic cooling systems. However a few hydronic cooling systems are of an open type in which the water comes into direct contact with air and can pick up unlimited quantities of oxygen so that galvanized pipe, or non-ferrous piping, pumps, and control valves are required to prevent rust formation. Black iron pipe with insulation which does not absorb moisture or has a vapor barrier is usually used with closed hydronic cooling systems.

Hydronic heating systems which produce winter comfort may not promote active air circulation. Floor, wall, or ceiling panel heating systems produce comfort conditions with no forced air circulation. Since circulation of air at lower than body temperature and at low humidity produces a cooling effect, winter comfort temperatures may be in the 70 to 72° range with radiant heating, while hot air heating systems which provide air circulation may require temperatures of 74 to 76° for the same degree of winter comfort.

If panel cooling were to be used without air circulation lower summer temperatures would be required than with air circulating systems, which in this case would be a disadvantage. Some work has been done in panel cooling but it must be used in conjunction with some air drying equipment so that the dew point of the air is kept well below the panel temperatures to prevent condensation on the panels. In comfort cooling systems forced air circulation is used as a means of carrying heat and moisture out of the conditioned space.

Air Conditioning Units

The air conditioning equipment which regulates air temperature and humidity, and provides air circulation varies considerably in its application and components. "Central station" units may provide all air treatment required for one or many "zones" in a building and may require all of the capacity of one or more chillers from 2 or 3 tons capacity in a home up to 100 tons or more in a commercial building. Buildings broken up into many rooms such as hotels, motels, apartment houses, schools, etc. may use individual air conditioning units so that each room or apartment constitutes a "zone" with its individual control. Any number of these zone units may be connected into and be a part of one hydronic cooling system which includes one or more water chillers.

For "Central station" units one supply pipe and one return pipe are all that is required to circulate the chilled water. Duplicate piping may be used to supply heating or the same piping can be used to provide summer cooling and winter heating. Typical central station units are described in the following paragraphs.

Fig. 2 shows an air conditioner cabinet which does not use a surface type heat exchanger. It is a combination air washer, cooler, and dehumidifier. Fresh air and return air mix and pass through a chilled water spray section. The chilled water spray, being below the dew point of the air passing through it, cools, dehumidifies, and washes the air. Eliminator sections down-stream from the spray, separates the entrained moisture (fine drops) from the air. The final air dew point and temperature can be controlled by varying the water temperature to the sprays and the return air by by-pass dampers.

This type of system is an open system and the water coming into direct contact with the air will pick up oxygen which could cause rust in iron piping and parts. Non-corrosive type piping and pumps should be used where this type of air chiller is used.

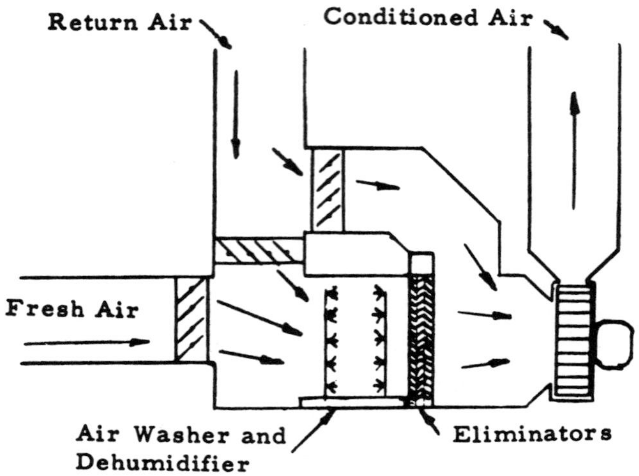

Fig. 2 — Air conditioner with chilled water supply to cool the air.

Fig. 3 shows a two zone air conditioner cabinet using one surface type heat exchanger. For summer time use, the return and fresh air dampers may be thermostatically operated to provide a minimum of fresh air when outside temperature is higher than return air. For conditions where the fresh air may be cooler than return air the dampers would provide maximum fresh air. Where humidity control is desired the dew point leaving the coil may be controlled by the chilled water temperature and some reheat can then be used to provide temperature and relative humidity control especially in spring and fall.

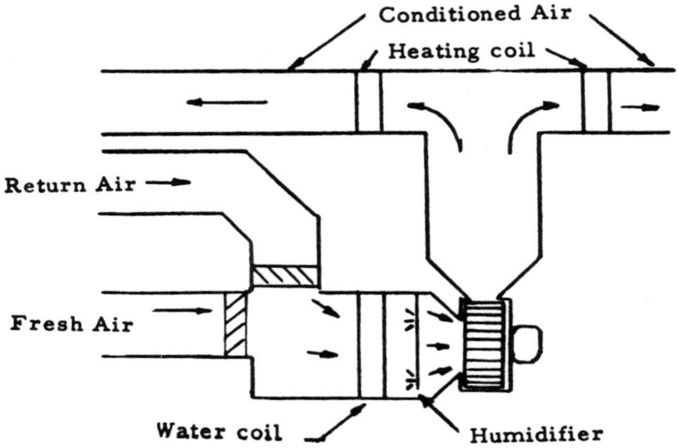

Fig. 3 — Two zone air conditioner with one surface type heat exchange.

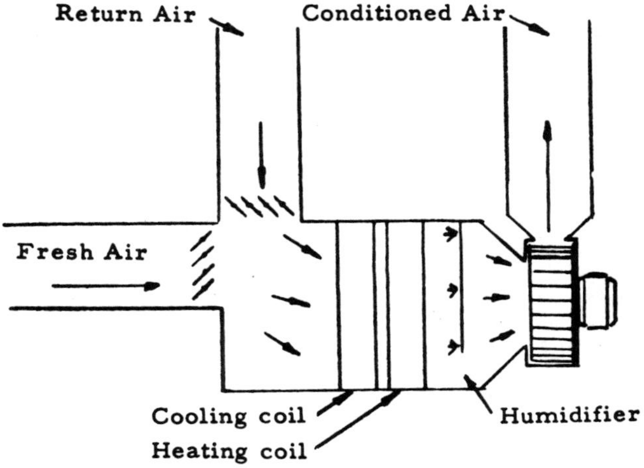

Fig. 4 — Reheat system employs a chiller water coil and a hot water coil.

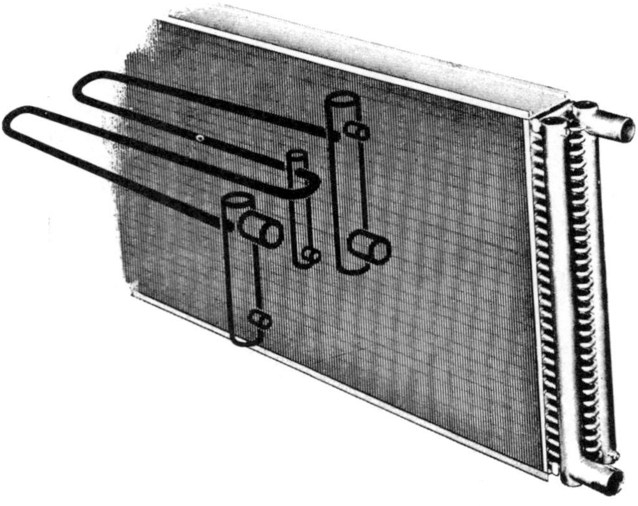

Fig. 5 — Water coil details showing inlet and outlet manifolds and parallel flow between headers.

Fig. 4 shows a chilled water coil and hot water coil arranged so that cooling and reheat can be used in spring and fall to control humidity. In hot summer weather the cooling coil only may be used while only the heating coil may be used in the winter season. Where necessary a humidistat may be employed to control relative humidity.

Fig. 5 shows the details of the water coil itself. It is provided with inlet and outlet manifolds and parallel flow of water between headers. In piping such a coil it is important that a counter flow piping arrangement should be provided.

The heat transfer rate in an air conditioning coil with all other conditions being equal, varies directly as the mean temperature difference (mtd) between the air over the coil and the water through it if only sensible heat is being transferred. When chilled water below the dew point of the air is circulated through the coil, moisture condenses on the fin and tube surface and the heat transfer rate increases. The number which indicates the ratio of the heat transferred with condensation forming on the coil, to that transferred with a dry coil is called the "wetted surface factor" (wsf). The total heat transfer in a coil removing both sensible and latent heat varies as the product of the mtd and wsf.

The mtd is calculated from:

$$\text{mtd} = \frac{\text{ltd} - \text{std}}{\log_e \frac{\text{ltd}}{\text{std}}}$$

or may be selected from a prepared table. In the above formula ltd stands for "large temperature difference" and std stands for the "small temperature difference" and the two of them represent the difference between air and water temperatures at the coil inlet and outlet.

Fig. 6A is a sketch showing parallel flow with the incoming chilled water entering in heat exchange relationship with the incoming air and the leaving water in heat exchange relationship with the leaving air. Fig. 6A shows that with parallel flow the mean temperature difference for the water and air temperatures shown is 24.8°.

If the incoming chilled water is in heat exchange relationship with the leaving air and the leaving water is in heat exchange relationship to the entering air as shown in Fig. 6B the mean temperature difference with the same inlet and outlet air and water temperatures is increased to 28.1° mtd. This means that the coil capacity will be 28.1-24.8 or 13.3% more when using counter flow instead of parallel flow with the air and water temperatures shown and when considering only the effect of mtd.

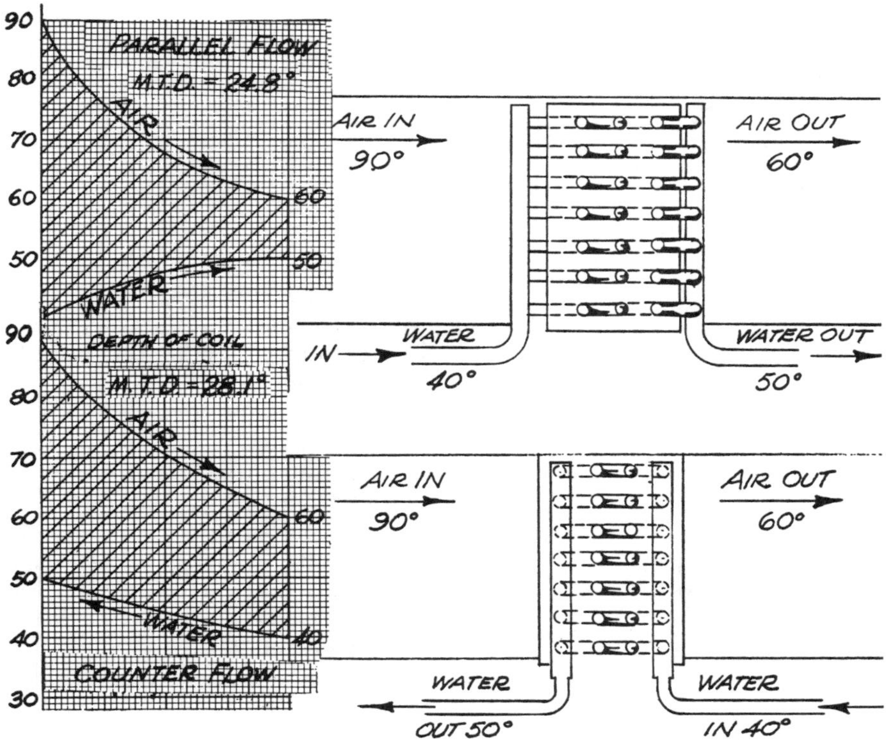

Fig. 6A and Fig. 6B

The piping for a closed hydronic cooling (and heating) system where all of the water from the chiller goes to one air cooling unit is a simple two pipe system with the supply and return pipes of the same diameter throughout. Manually operated vent valves should be provided at high points in the pipe to allow air to be removed from the piping as it is being filled. Air removal equipment and an expansion tank should be provided.

If the system is to be used for hydronic cooling only, one expansion tank volume of 3% of the water volume is ample. If the system is to provide cooling and heating through the same piping the expansion tank volume should be about 22% of system volume for water temperatures up to 250°F. in heating. A water pressure reducing valve to feed water into the piping should be set to maintain a pressure of about 4

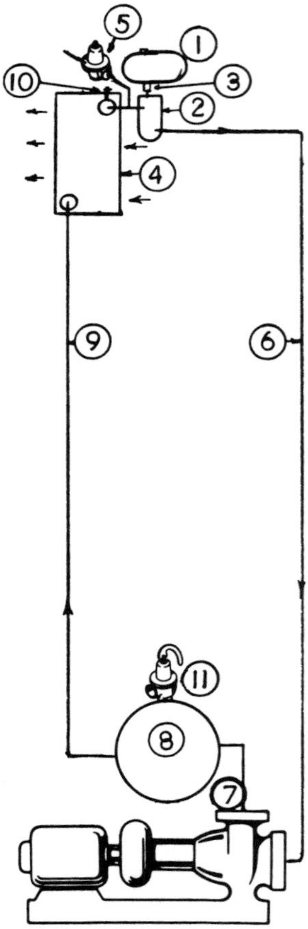

Fig. 7 — Components of system used in high rise buildings:

1. Expansion tank—volume 3% of system water volume.
2. Air separator—removes bubbles from water.
3. Airtrol fitting—removes excess air from tank.
4. Air cooling coil.
5. Pressure reducing check valve —adds water if system pressure goes below 4 psi.
6. Return chilled water pipe.
7. Chilled water pump.
8. Water chiller.
9. Chilled water supply pipe.
10. Manual air vent.
11. Pressure relief valve

psi at the high point. With proper pipe sizing this will insure against any possible cavitation at the circulating pump.

The circulating pump should normally have its suction side tied in to the expansion tank and receive its water from the outlet of the air separator. However, in high rise buildings such as shown in Fig. 7, it may be practical to locate the air separator and expansion tank at or near the air-conditioning cabinet at the high level. The ideal place for the air separator is at the point where the water is at the highest temperature and lowest pressure. The pump can be located in the machinery room along with the chiller since, in this example, there is no danger of cavitation at the pump.

Systems which have a large number of coils cooled by water from one package liquid chiller are sketched and described briefly in Figs. 7A, 7B, 7C, 7D, and 7E.

Systems A, and B maintain constant chilled water flow rate through the system and the individual zone cooling coils. Zone control can be accomplished individually by using zone thermostats to stop and start fan motors or by the modulation of face and by-pass dampers regulating air flow over the coils.

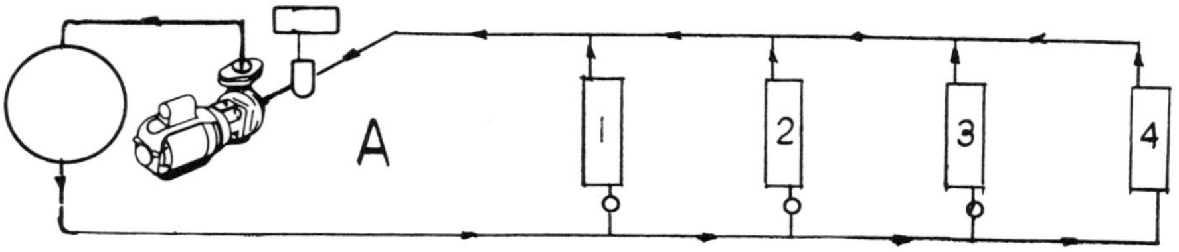

Fig. 7A.—Direct return system has the shortest supply piping and shortest return piping to No. 1 coil and requires balancing valves to adjust flow to equal that through the most distant coil No. 4 as water resistance is greater in both supply and return lines from No. 4 coil.

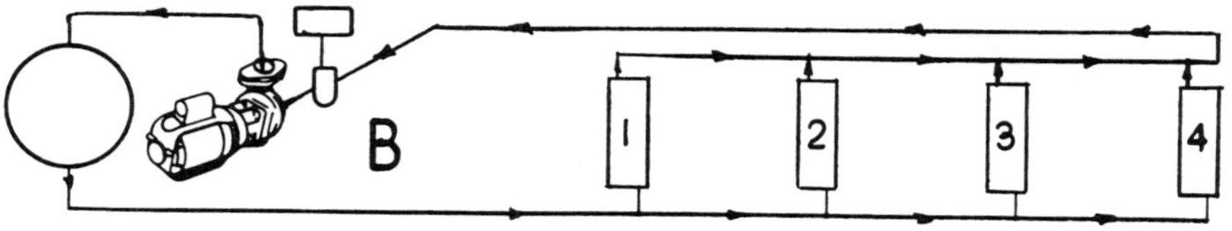

Fig. 7B—Reverse return system has the same total length of supply and return piping from the chiller to each coil. No. 1 coil has the shortest, supply and longest return while No. 4 coil has the longest supply and shortest return. This system will tend to have equal flow through all coils of the same characteristics without the use of balancing valves.

System C uses a constant air flow over the coils and the water flow through each zone coil is started or stopped as the zone thermostat stops and starts the individual secondary pumps.

System D may use constant air flow while individual zone control valves modulate or stop and start water flow through the zone coils. The loaded check valve by-passes the water which passes through the coils when all zones are working at full capacity. The system pump

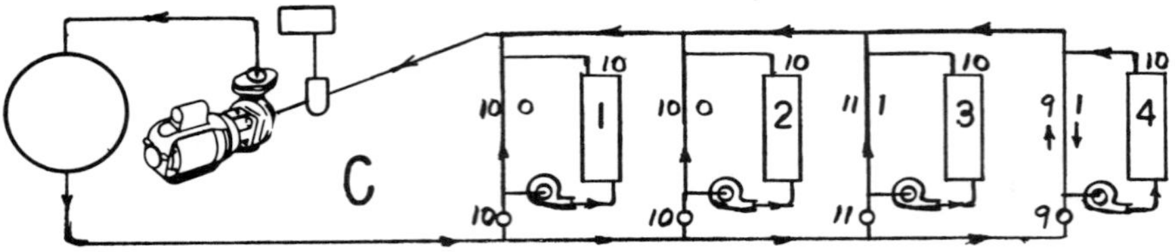

Fig. 7C—Primary secondary pumping system with cross over pipes from the supply pipe to the return pipe uses balancing valves to regulate the flow at each coil from the supply to the return piping. The sketch shows one primary pump which circulates 40 gpm through the system. All secondary pumps circulate 10 gpm through their individual coils. As secondary pumps are "on" due to the zone thermostat call for cooling all water through the balancing valves No. 1 and 2 is circulated through the respective coils. As the individual thermostats indicate, cooling demand is satisfied in their zones the pumps stop and the water through the balancing valves returns through the by pass pipes and return lines. No. 3 balancing valve is shown as allowing an 11 gpm flow through No. 3 cross over line. When the pump is on, 10 gpm passes through the pump and coil and one gpm goes through the cross over pipe which flows 11 gpm with the pump "off". In No. 4 with 9 gpm through the balancing valve 10 gpm is pumped through the coil when the pump is "on" including one gpm of water returning through the cross over pipe.

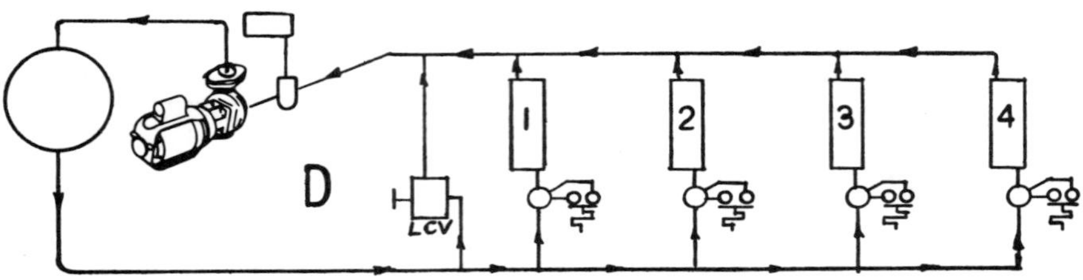

Fig. 7D—Thermostatic zone control and loaded check valve system has a loaded check valve which by-passes water from the supply line to the return line whenever the zone valves modulate or close off. The loaded check valve insures essentially a constant pressure drop between supply and return lines sufficient to supply the required flow through each coil when the zone control valves open.

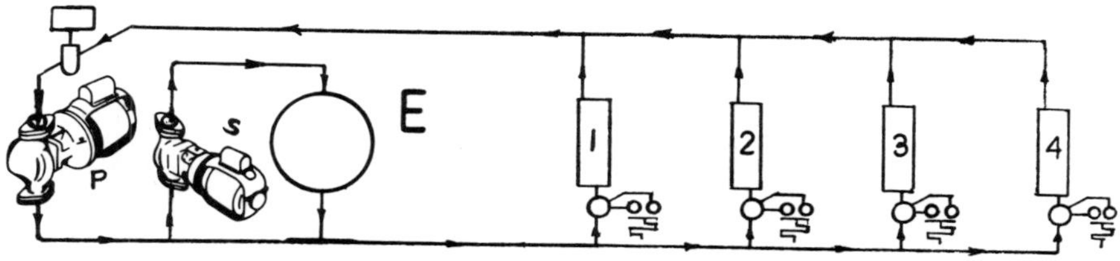

Fig. 7E—Modulated system flow rate and zone control valves. In this system the primary pump (P) pumps the water through the system. As zone control valves close, the flow rate in the system decreases and static head rises. The necessary constant flow rate through the chiller is maintained by the secondary pump (S) independently of the primary flow.

should be shut down when the chiller operation stops. A minimum flow must be maintained while the pump operates to prevent pump overheating.

System E has a constant flow rate through the chiller while the zone valves modulate the system flow either by one at a time shut off or modulating flow through the coils. Air control equipment is provided ahead of the system pump.

Pumps and Pumping

In carrying heat from one point to another hydronically, water acts as a heat container on a conveyor, which fills up with heat to a higher level at one point and dumps that heat at another point along the conveyor. The conveyor in this case is the pipe and the pounds of water, the heat containers, passing through the pipe. The pump supplies the energy to move the heat containers or water through the pipe.

The required pump characteristics are determined by the rate of water flow and the difference in head through which the water is pumped. Fig. 8 shows two tanks of water with the level in one tank 10 ft. above that of the other. Pump A is pumping water horizontally from one tank to the other. Pump B is pumping water from the bottom of the lower tank to the upper part of the higher tank. In pump A the suction head is 5 ft. and the discharge head is 15 ft. The pump head = 15 ft. – 5 ft. = 10 ft. Pump B has a 15 ft. suction head and 25 ft. discharge head so it is pumping against 25 ft. – 15 ft. or 10 ft., the same as pump A. If pump A was pumping 100 pounds of water per minute the work being done by it would be 100 lb. per min x 10 ft. head = 1000 ft. lbs. per min.

If pump B was pumping 33 pounds of water per minute it would be doing 33 lb. per min. x 10 ft. = 330 ft. lbs. per minute.

Since one horse power is defined as 33,000 ft. lbs. per min., pump B output would be

$$\frac{330 \text{ ft. lbs. per min.}}{33{,}000 \text{ ft. lbs. per hp min.}} = 0.1 \text{ hp.}$$

If the efficiency of the pump was 50% the power required to drive it would be:

$$\frac{0.1 \text{ hp}}{.50} = .02 \text{ hp} = 1/50 \text{ hp Motor Required.}$$

Power is the rate of doing work and this can be measured (for a pump) in terms of pounds of material pumped per minute, multiplied by the difference in elevation through which it is moved. Thus pounds per minute x feet = foot pounds per minute.

Where the rate of flow is specified in gallons per minute (1 gpm = 231 cubic inches per minute), it is necessary to multiply the volume flow rate by the weight per unit of volume to get the flow rate in pounds per minute.

Thus, gpm x pounds per gallon = pounds per minute, or cfm (cubic feet per minute) x pounds per cubic foot = pounds per minute.

Water at normal temperatures has a density of about 62.4 pounds per cubic foot (1728 cubic inches), or $\frac{231}{1728} \times 62.4 = 8\text{-}1/3$ pounds per gallon, so that the horsepower in terms of gpm and feet of head is:

$$\text{hp} = \frac{\text{gpm x 8-1/3 lbs. per gal. x feet of head (increase in elevation)}}{\text{33,000 foot pounds per horsepower minute}}$$

$$\text{hp} = \frac{\text{gpm x feet of head}}{3960}$$

The horsepower as indicated above is the power developed by the pump and the input to the pump is:

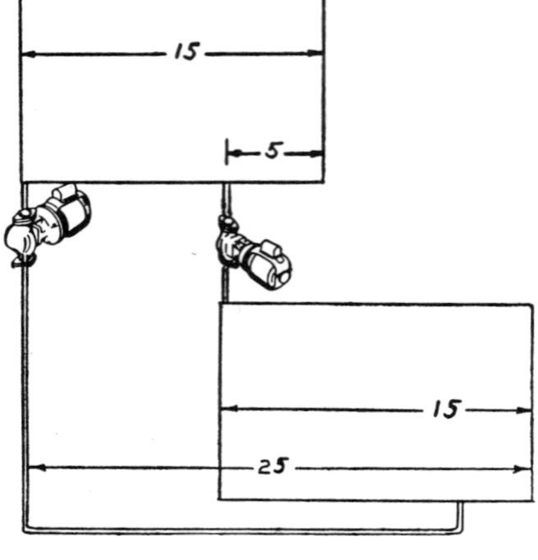

Fig. 8

$$\text{Input hp} = \frac{\text{gpm x head (in feet)}}{3960 \text{ x m.e}}$$

where m.e is the mechanical efficiency of the pump. At high temperature, water weighs less than 62.4 pounds per cubic foot, or 8-1/3 pounds per gallon, so the pump horsepower becomes less in proportion to the change in pounds per cubic feet.

Power is also a product of the volume pumped per minute times the difference in pressure through which it is pumped. At normal temperatures a column of water 27.72 inches, or 2.31 feet in height, exerts a

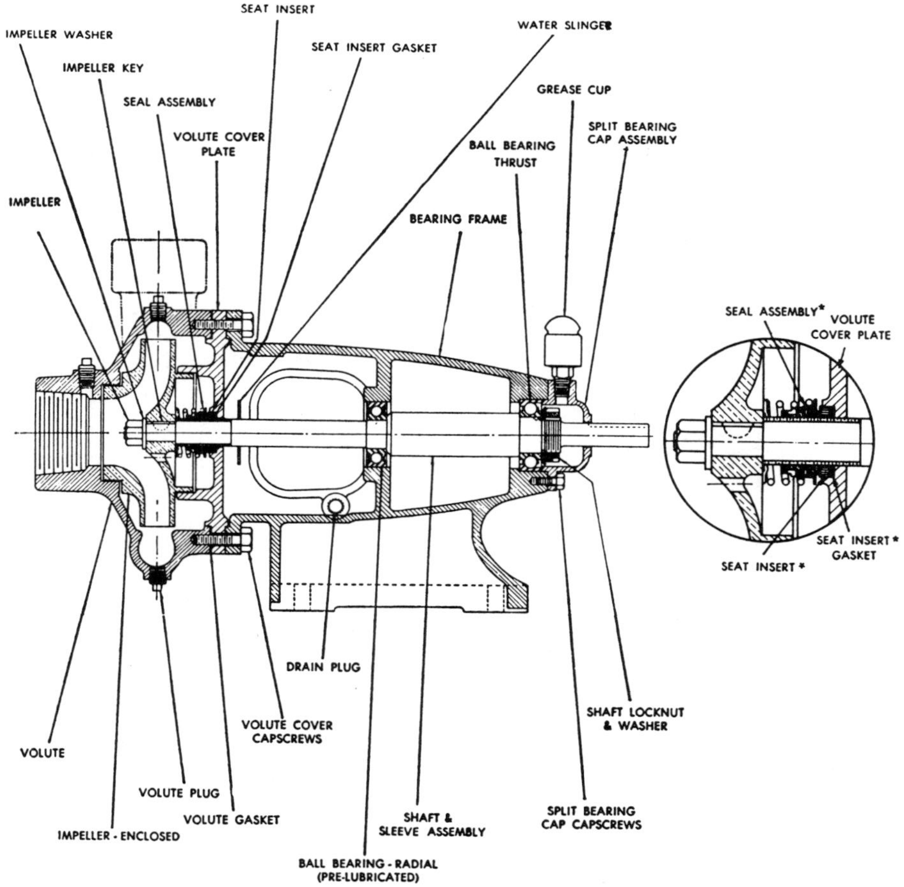

Fig. 9

pressure on the bottom of the column of one pound per square inch so that the horsepower formula becomes:

$$hp = \frac{gpm \times psi\ (\text{pounds per square inch}) \times 2.31\ \text{ft. per psi}}{3960 \times m.e}$$

$$\text{or } hp = \frac{gpm \times psi\ (\text{pressure difference})}{1714 \times m.e}$$

This latter formula does not depend on the density of the water because as the pounds of water decreases per gallon, the head per psi increases so that the two factors counterbalance each other, and no error occurs due to reduced liquid density at high temperature.

Fig. 9 shows a cross section of a typical centrifugal pump which can be driven at motor speed by using a flexible coupling from the motor shaft to the pump shaft. It could also have a drive pulley mounted on the shaft and be driven by a V belt. In the drawing, the water will enter in a horizontal suction pipe threaded into the pump inlet. It enters the impeller and is thrown out of the impeller by backward curved vanes in all directions. The volute casing has a water passage increasing to maximum cross sections at the outlet and is designed to maintain a constant velocity as it collects the water from all parts of the wheel periphery. A shaft seal is used to prevent water leakage where the impeller shaft enters the casing.

Fig. 10 shows the general characteristics of a centrifugal pump and shows the pump head (difference between outlet and inlet heads in feet of water), efficiency, and power input, plotted against the flow rate

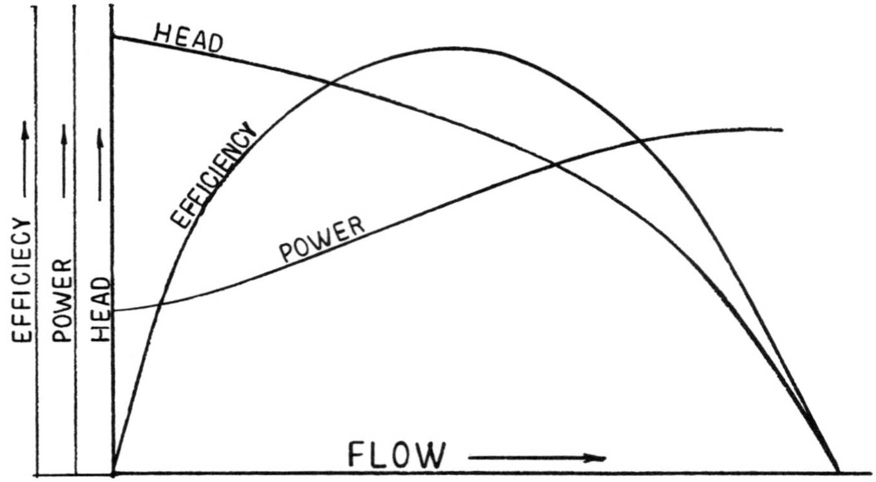

Fig. 10

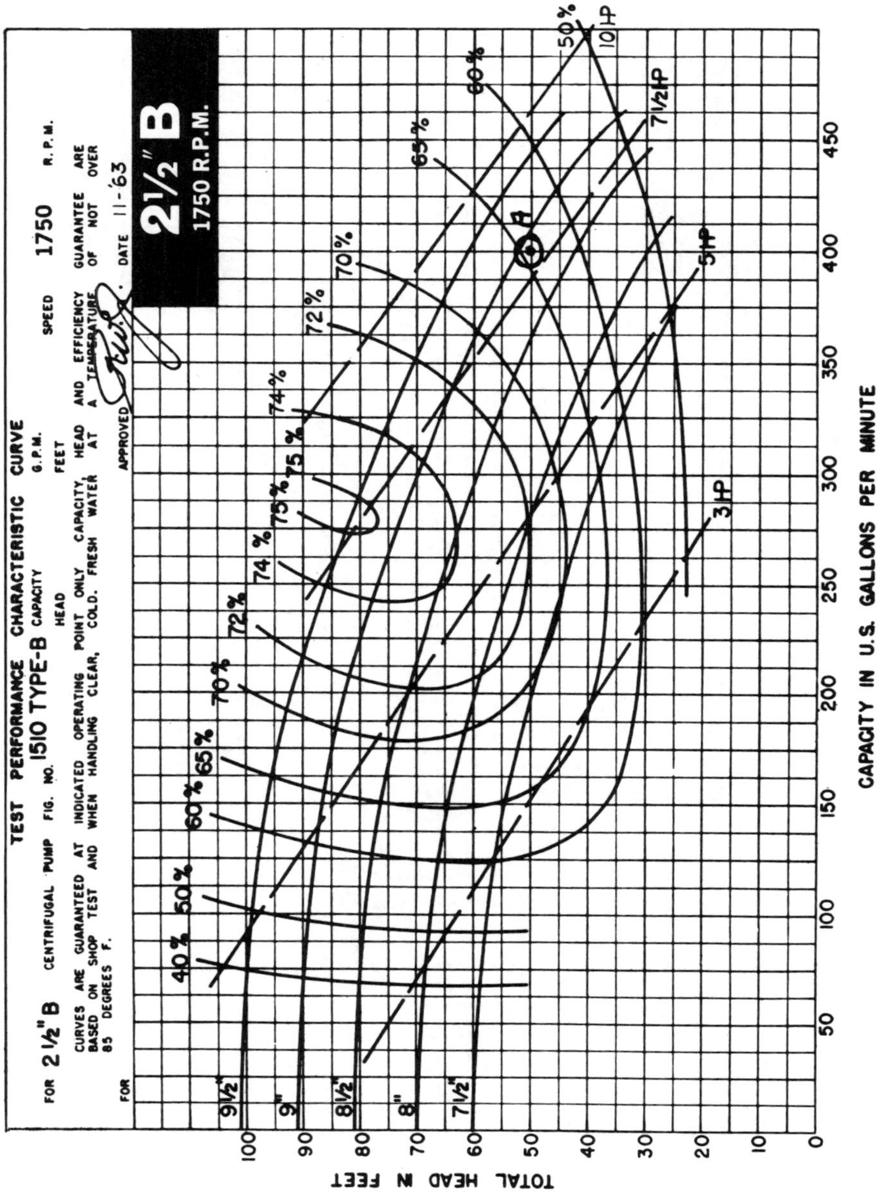
TEST PERFORMANCE CHARACTERISTIC CURVE
FOR 2½" B CENTRIFUGAL PUMP FIG. NO. 1510 TYPE-B CAPACITY G.P.M. SPEED 1750 R.P.M.
HEAD FEET
CURVES ARE GUARANTEED AT INDICATED OPERATING POINT ONLY. CAPACITY, HEAD AND EFFICIENCY GUARANTEE ARE BASED ON SHOP TEST AND WHEN HANDLING CLEAR, COLD, FRESH WATER AT A TEMPERATURE OF NOT OVER 85 DEGREES F.
FOR
APPROVED
DATE 11-'63
2½" B
1750 R.P.M.
TOTAL HEAD IN FEET
CAPACITY IN U.S. GALLONS PER MINUTE
9½"
9"
8½"
8"
7½"
3HP
5HP
7½HP
10HP
A

Fig. 11A

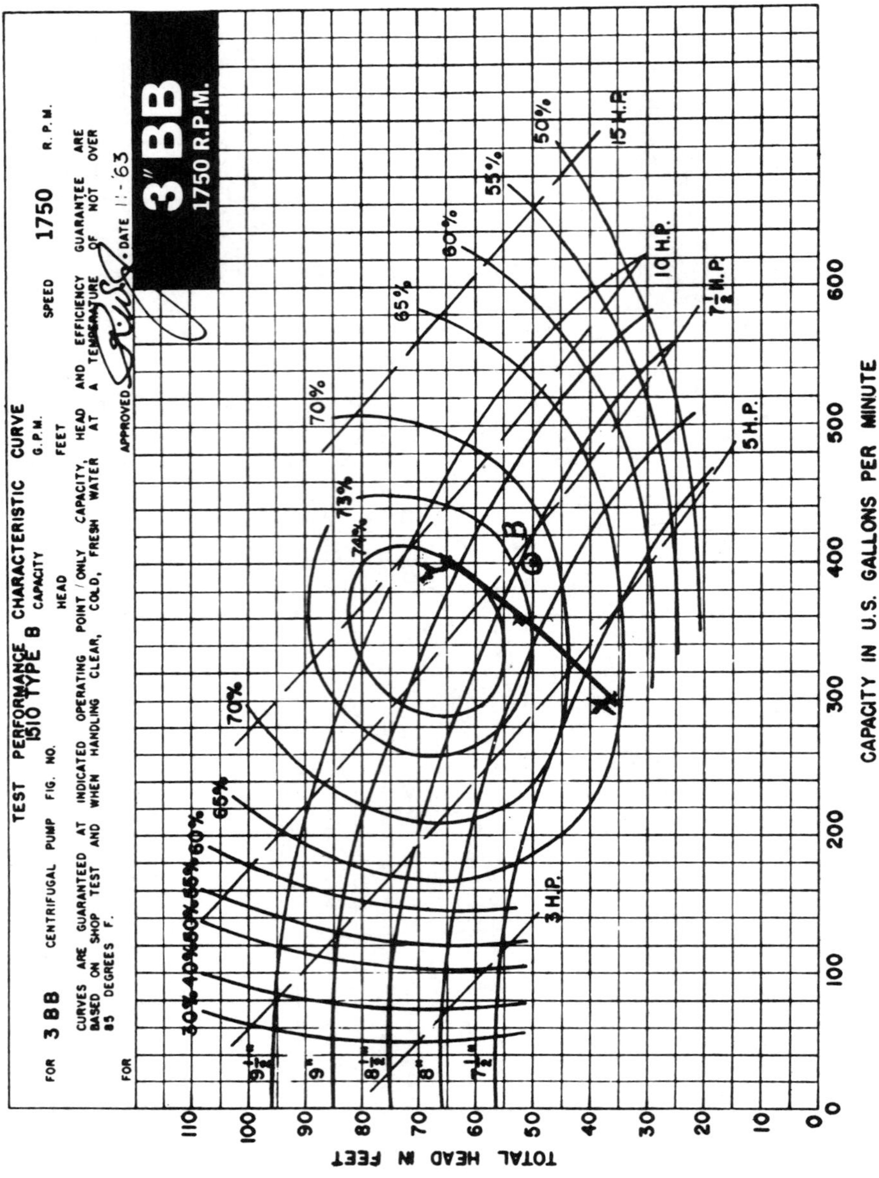
TEST PERFORMANCE CHARACTERISTIC CURVE
FOR 3 BB CENTRIFUGAL PUMP FIG. NO. 1510 TYPE B CAPACITY G.P.M.
HEAD FEET
SPEED 1750 R.P.M.
CURVES ARE GUARANTEED AT INDICATED OPERATING POINT / ONLY CAPACITY, HEAD AND EFFICIENCY GUARANTEE ARE BASED ON SHOP TEST AND WHEN HANDLING CLEAR, COLD, FRESH WATER AT A TEMPERATURE OF NOT OVER 85 DEGREES F.
FOR
APPROVED
DATE
3" BB
1750 R.P.M.
CAPACITY IN U.S. GALLONS PER MINUTE
TOTAL HEAD IN FEET
3 H.P.
5 H.P.
10 H.P.
15 H.P.
70%
73%
74%
65%
60%
55%
50%

Fig. 11B

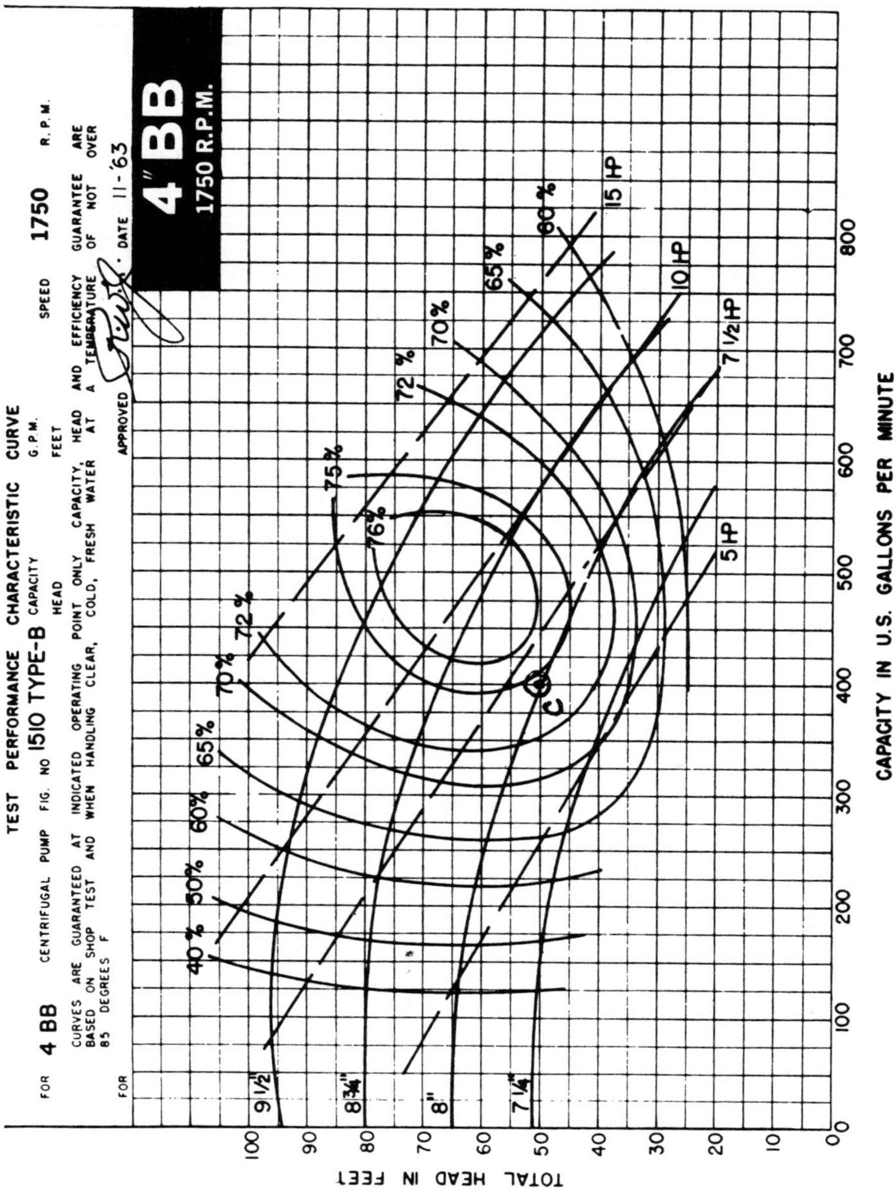
TEST PERFORMANCE CHARACTERISTIC CURVE
FOR 4 BB CENTRIFUGAL PUMP FIG. NO 1510 TYPE-B CAPACITY G.P.M. HEAD FEET SPEED 1750 R.P.M.
CURVES ARE GUARANTEED AT INDICATED OPERATING POINT ONLY CAPACITY, HEAD AND EFFICIENCY GUARANTEE ARE BASED ON SHOP TEST AND WHEN HANDLING CLEAR, COLD, FRESH WATER AT A TEMPERATURE OF NOT OVER 85 DEGREES F
FOR
APPROVED
DATE 11-'63
4" BB
1750 R.P.M.
TOTAL HEAD IN FEET
CAPACITY IN U.S. GALLONS PER MINUTE
9 1/2"
8 3/4"
8"
7 1/4"
40 %
50%
60%
65%
70%
72 %
75%
76%
80 %
5 HP
7 1/2 HP
10 HP
15 HP
C

Fig. 11C

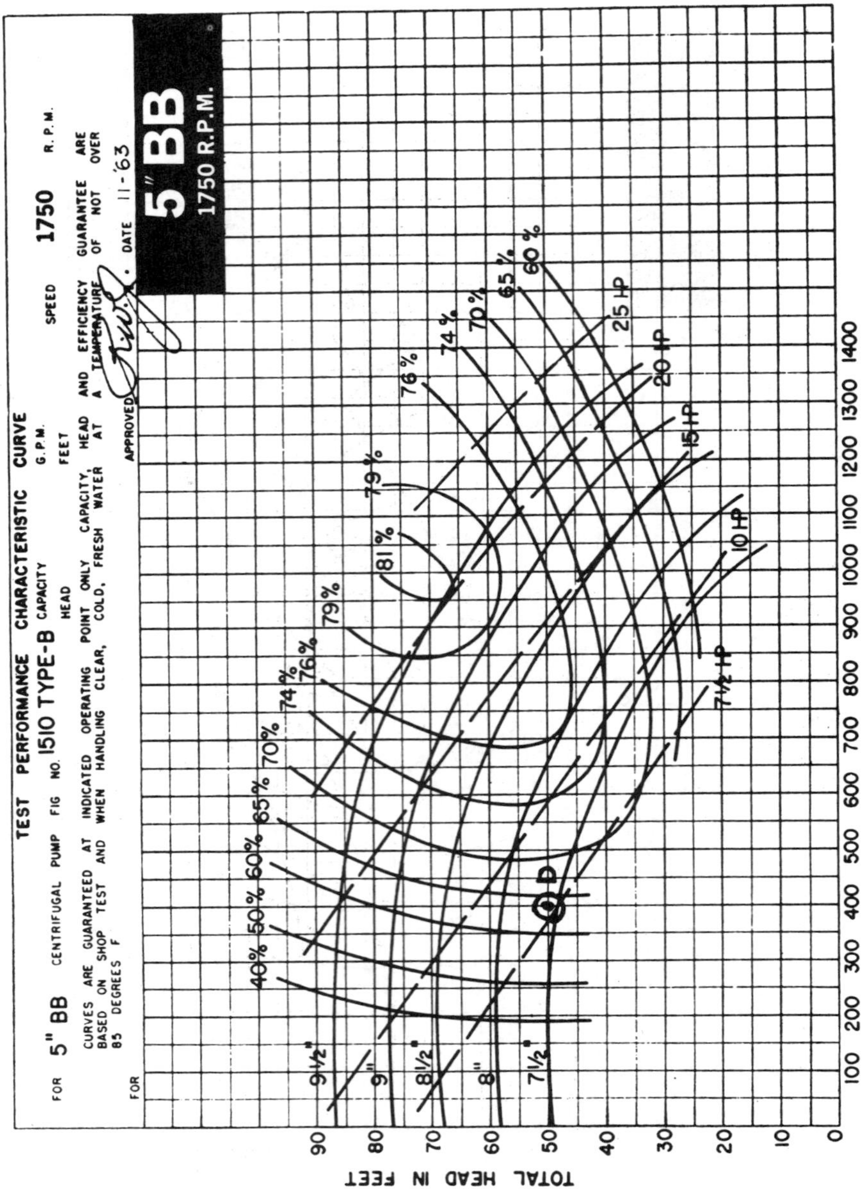
TEST PERFORMANCE CHARACTERISTIC CURVE
FOR 5" BB CENTRIFUGAL PUMP FIG NO. 1510 TYPE-B CAPACITY G.P.M. HEAD FEET SPEED 1750 R.P.M.
CURVES ARE GUARANTEED AT INDICATED OPERATING POINT ONLY CAPACITY, HEAD AND EFFICIENCY GUARANTEE ARE BASED ON SHOP TEST AND WHEN HANDLING CLEAR, COLD, FRESH WATER AT A TEMPERATURE OF NOT OVER 85 DEGREES F
FOR
APPROVED
DATE 11-'63
5" BB
1750 R.P.M.
40% 50% 60% 65% 70% 74% 76% 79% 81% 79% 76% 74% 70% 65% 60%
25 HP
20 HP
15 HP
10 HP
7½ HP
9½"
9"
8½"
8"
7½"
D
CAPACITY IN U.S. GALLONS PER MINUTE
TOTAL HEAD IN FEET

Fig. 11D

through the pump. Fig. 11A, 11B, 11C, and 11D are performance curves for Bell & Gossett series 1510 pumps operating at 1750 rpm and in sizes of 2½", 3", 4" and 5" respectively.

If the pumping load for a particular system is approximately 400 gpm at a head of 50 feet it is noted as point A on the 2½ inch pump performance curve. A 9 inch curve is slightly above this point and an 8½ inch impeller curve is slightly below the point. For this pump casing the impeller would have to be a little less than 9 inch diameter for exactly 400 gpm at 50 ft. head or the 9 inch diameter impeller would deliver a little more than the 400 gpm specified. Point A is between the 60% and 65% efficiency curves so the efficiency will be about 64%. The point A is above and to the right of the dashed line labelled 7½ hp so will use a little more than 7½ hp.

If the 3 inch pump at point "B" on the 3 inch pump performance sheet was used it will be noted that the impeller of 8½ inch diameter will give a little over 400 gpm flow rate, the efficiency is above 71.5% and the horsepower is less than 7½.

The 4 inch pump at point "C" will exactly fill the capacity and head requirement with an 8 inch impeller at an efficiency of 74.5% and use less than 7½ hp.

The 5 inch pump at point "D" will almost handle the flow required with a 7½ inch impeller and will require a little over 7½ hp at an efficiency of about 63 or 64%.

It is evident then that the 2½ inch pump is a little too small and the 5 inch pump is too big. Either the 3 or 4 inch pumps could be used and would underload a 7½ hp motor.

Pressure Losses

The pressure losses in a chilled water circuit that have to be overcome by the pump are the friction losses in the chiller, the friction losses in the air cooling coil and the losses in the piping used to carry the water from one heat exchanger to the other and return. If there are any control valves in the system which will restrict flow and cause a pressure drop, that loss will also be included in the total pressure losses.

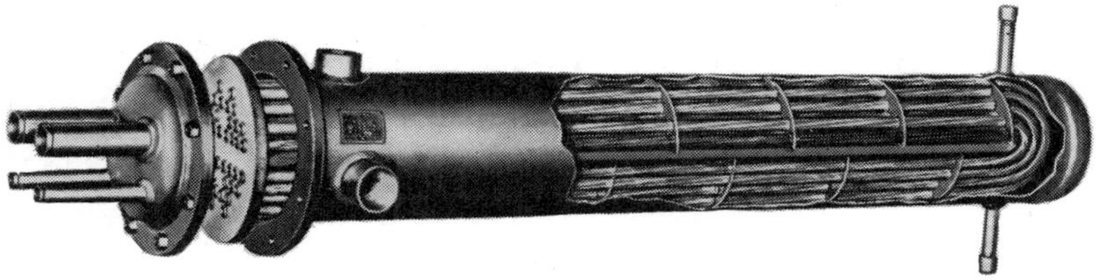

Fig. 12

Rating Sheet For 75 hp Liquid Chiller

CHILLED WATER TEMPERATURES		85°F. CONDENSER WATER 105°F. CONDENSING TEMPERATURE						
Entering Chiller °F.	Leaving Chiller °F.	Capacity Tons	BTU/Hour	Bhp.	Chilled Water GPM	PD Thru Chiller (Ft.)	Condenser Water GPM	PD Thru Condenser (Ft.)
46	40	66.9	802,800	78.4	268	9.71	*207	18.71
48	40	66.9	802,800	78.4	201	5.54	*207	18.71
50	40	66.9	802,800	78.4	161	3.70	*207	18.71
48	42	69.7	836,400	79.3	279	10.71	202	2.92
50	42	69.7	836,400	79.3	209	5.94	202	2.92
52	42	69.7	836,400	79.3	167	3.98	202	2.92
50	44	72.7	872,400	80.0	291	11.32	217	3.38
52	44	72.7	872,400	80.0	218	6.24	217	3.38
54	44	72.7	872,400	80.0	174	4.35	217	3.38
51	45	73.9	886,800	80.3	296	11.55	225	3.65
53	45	73.9	886,800	80.3	222	6.47	225	3.65
55	45	73.9	886,800	80.3	177	4.45	225	3.65
52	46	75.2	902,400	80.6	301	11.90	235	3.95
54	46	75.2	902,400	80.6	226	7.00	235	3.95
56	46	75.2	902,400	80.6	180	4.62	235	3.95
54	48	78.3	939,600	81.3	313	12.82	256	4.70
56	48	78.3	939,600	81.3	235	7.50	256	4.70
58	48	78.3	939,600	81.3	188	4.90	256	4.70
56	50	81.3	975,600	81.8	325	13.80	279	5.56
58	50	81.3	975,600	81.8	244	8.03	279	5.56
60	50	81.3	975,600	81.8	195	5.24	279	5.56

Fig. 13

Chiller Losses

The pressure losses at various flow rates through the liquid chiller depend upon the design of the chiller itself. Fig. 12 shows a liquid chiller construction. The water entering the shell at the top makes several passes across the top half of the tube bundle as it passes from the front to the back and an equal number of passes in the bottom half as the water works forward to the outlet. The baffle spacing or number of baffles determines the number of passes across the tubes and the water friction losses increase as the number of passes increase. The actual friction losses for any manufacturers chiller should be given in the rating sheets.

Fig. 13 is rating sheet for a 75 horsepower package liquid chiller. The chilled water flow rates are based on 6, 8, or 10F td. (temperature difference) of the water through the chiller or on water flow rates of 4 gpm per ton, 3 gpm per ton and 2.4 gpm per ton of actual refrigeration capacity.

For an 8F temperature reduction in the chiller the flow rate is shown (with 105F condensing temperature) as 218 gpm and a pressure loss of 6.24 ft. of head. This is based on 52F return water to the chiller and 44F supply water leaving the chiller. The highest water flow rate is shown as 307 gpm and a pressure loss of 12.29 feet. In general the friction loss through a pipe, chiller, or air cooling coil varies as the square of the rate of flow through them. In this case the pressure loss at 218 gpm flow rate is 6.24 ft. If the loss varies as the square of the flow rate the loss at 307 gpm would be: $\left[\frac{307}{218}\right]^2$ x 6.24 ft. = $(1.408)^2$ x 6.24 = 1.983 x 6.24 = 12.37 or very nearly equal to the 12.29 indicated in the table. Consequently if the head loss is known for any flow rate the head loss for any other flow rate can be determined as follows: $PD_2 = PD_1\ (gpm_2 - gpm_1)^2$

Coil Pressure Losses

The water pressure losses through the air cooling coil will depend upon the size of the water tubes used, the water velocity through the tubes, the length of the coil and the number of passes the water makes from the inlet header to the outlet header. The manufacturer of the air to water heat exchanger should indicate pressure drop through the coils or a method of coil selection which will include water pressure loss in the coil. Water velocities are usually maintained within a range of 2 feet per second to 5 feet per second. In the process of working out a proper selection of coil, fin area, tube lengths, and the number of tubes in parallel, enter in to assure the coil will handle the design load at design water temperatures and produce the design ratio of sensible to total cooling.

In the selection of water chilled air conditioning coils a chart is worked out which indicates the water velocity and coil header loss based on the number of tubes in parallel and the rate of water flow. Fig. 14 shows this chart. Based on a water flow of 70 gpm and 30 tubes in parallel, and the size tubes used, the velocity in the tubes is indicated as about 4 feet per second.

The coil design or model has previously been worked out in coil selection calculations to be a 2,018 design. The velocity point is connected to the intersection of the model number horizontal line and line D. A straight line carried to line E indicates the water pressure loss per row of tubes for the various tube arrangements.

When each tube from the coil inlet manifold is passed through the downstream row of tubes and then proceeds through each row (by means of return bends) and to the return manifold the coil is of a single serpentine type and the pressure drop is given in column "S" as 2.1 feet per row. Chilled water coils are usually 4, 6, or 8 rows deep and if the pressure drop is based on 6 rows deep the water pressure drop through the coil will be as follows:

1.	Through tubes – 6 rows x 2.1 ft. per row =	12.6 ft.
2.	Header pressure loss (from line c)	1.1 ft.
	Total through coil at 70 gpm	13.7 ft.

In general the pressure drop through the coils can be expected to lie between 5 feet and 20 feet of head. With pressure loss known for any coil selection at any flow rate the pressure loss at the flow rates can be approximated by taking them as proportional to the square of the flow rates.

Pumping and Piping

The friction losses in the piping required to circulate the water from the chiller to the air cooling coil and back, depend upon the size of the pipe, the length of the pipe and the number and types of fittings used. Anything in a piping system which causes water velocity to increase, decrease, or change direction causes a friction pressure loss in a piping system. Also in a piping system which contains more than one air chilling unit the water flow is split into various paths and changes in pipe size occur.

Pressure losses due to pipe friction can be measured in terms of pounds per square inch per 100 feet of pipe. A smaller unit of pressure loss is feet of head per 100 feet of pipe and a still smaller unit is in terms of mil-inches per foot. A mil-inch is .001 inches so that 1000 mil-inches equals one inch of head and 12,000 mil-inches equals one foot of head.

A piping water friction chart is shown in Fig. 15. The horizontal scale shows water flow rates from .1 gpm up to 10,000 gpm on a logarithmic scale. The vertical scale is also logarithmic and calibrated from one

Table 4 — Head Loss of Water Flowing Through Fittings Expressed in Equivalent Length of Straight Pipe of the Same Size

Fitting	*Size*	½	¾	1	1¼	1½	2	2½	3	4	6
90° std. ell		1.0	1.6	2.1	2.6	3.1	4.2	5.2	6.2	8.3	12
45° std. ell		.7	1.1	1.4	1.8	2.2	2.9	3.6	4.4	5.8	8.7
90° long-sweep		.5	.8	1.0	1.3	1.6	2.1	2.6	3.5	4.2	6.2
Return bend		1.1	1.7	2.3	2.9	3.4	4.6	5.7	6.9	9.2	12
Open gate valve		.5	.8	1.0	1.3	1.6	2.1	2.6	3.1	4.2	6.2
Open globe valve		12	19	25	32	37	50	62	75	100	150
Open cock stop		1.0	1.6	2.1	2.6	3.1	4.2	5.2	6.2	8.3	12
Tee-straight through		.5	.8	1.0	1.3	1.6	2.1	2.6	3.1	4.2	6.2
Tee- 25% to branch		16.0	25	33	42	50	67	83	100	133	200
Tee- 50% to branch		4.0	6.2	8.3	10	12	17	21	25	33	50
Tee-100% to branch		1.8	2.8	3.7	4.7	5.6	75	9.4	11	15	23

mil-inch per foot up to 5,000 mil-inches per foot. Lines slanting upward and to the right indicate pipe sizes from ¼ inch standard pipe to 12" nominal pipe size. Pipe classified as extra heavy or double extra heavy have the same outside diameters as standard pipe but have greater wall thickness and therefore smaller inside diameters and offer more restriction to flow than standard pipe. Galvanized and black piping of the same nominal pipe size offer the same resistance to flow.

The pressure loss per foot of pipe due to friction is an indication of the rate of flow through the pipe and at high velocity the water causes piping noise. Consequently pipe sizes can be selected to maintain acceptable noise levels by limiting pressure losses per foot of pipe to an acceptable level.

A general statement on piping noise can be made as follows:

Where the noise level must be kept low as for residence or office sound levels, the velocity should be kept down to 4 feet per second or less in pipes up to 1½ inch nominal. For larger pipes a limiting pressure loss of 500 mil-inches per foot or less should be maintained. Where the piping is used in commercial places or industrial applications where sound levels are normally high, higher velocities may be allowed. Excessive velocity may tend to erode pipe elbows or fittings which cause a change in direction to occur.

Equivalent Length of Pipe

For piping systems as used in many applications having a normal number of elbows and other fittings the equivalent length of pipe is often estimated as 1.5 times the actual pipe length. This means that the pressure loss through 100 feet of pipe and the normal number of fittings would be approximately the same as in 150 feet of straight pipe of the same size and at the same flow rate.

WATER VELOCITY AND PRESSURE DROP DETERMINATION

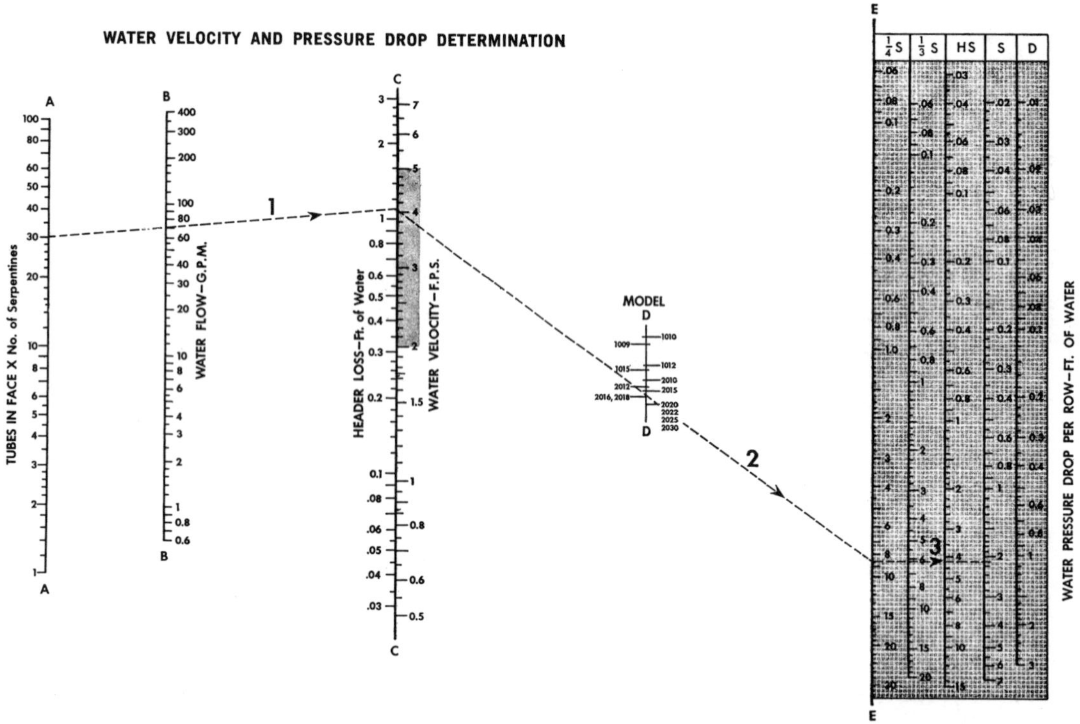

EXPLANATION

1. Connect, Tubes In Face × Number of Serpentines, on line A with Water Flow, GPM on line B and find intersection on line C.
2. On line C, read Water Velocity feet per second and Header Loss feet of water. Select a circuiting arrangement that produces a water velocity in the optimum range. Chilled water, 2 to 5 fps; hot water, 1 to 3 fps.
3. From this intersection, draw a line through the correct model number on line D, and find intersection on line E. Note that Models 2016 and 2018 use the same point on line D. Similarly, Models 2020, 2022, 2025 and 2030 use a common point on line D.
4. Transfer this point horizontally to the correct serpentine scale as indicated by the headings, and read Water Pressure Drop Per Row, feet of water.
5. Total Pressure Drop = (Number of Rows × Pressure Drop Per Row) + Header Loss.

Fig. 14

FRICTION IN BLACK IRON PIPE

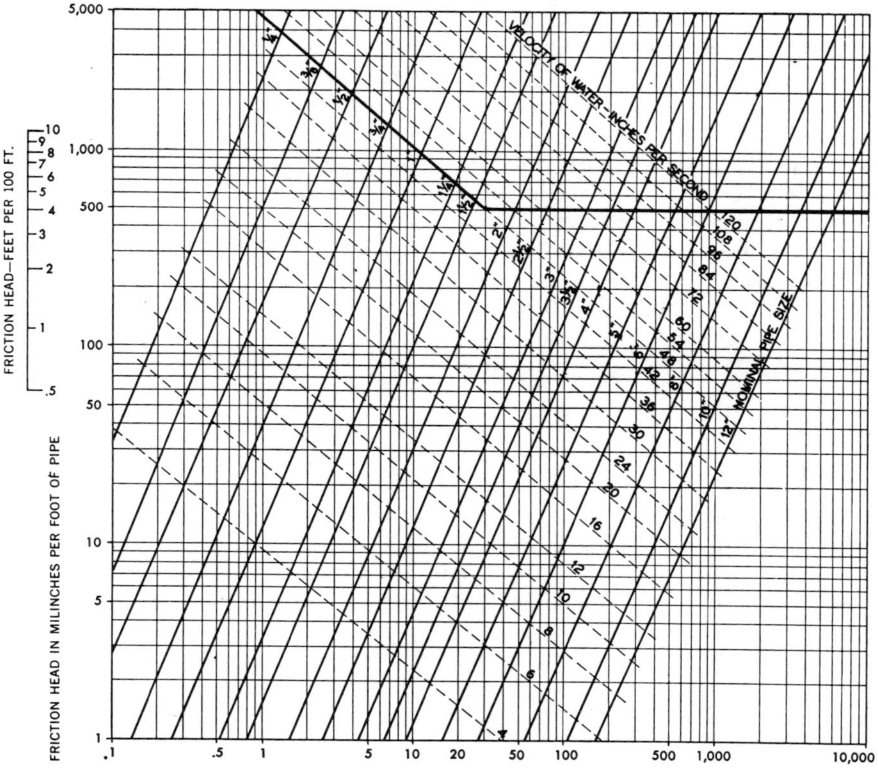

FLOW OF WATER IN GALLONS PER MINUTE

HOW TO USE THIS CHART

The above chart will enable the designer to accurately determine the milinch resistance and velocity occurring in pipes from ¼" to 12" at any given flow.

Example: When carrying a load of 20 GPM, what is the resistance and velocity in a 1½" pipe?

Reading up from the 20 GPM point at the bottom of the Chart to the intersection with 1½" pipe size, extend a line to the left border, where a resistance of 350 milinch per foot is indicated.

The velocity is indicated by the broken lines and is read at the intersection of the GPM line and the pipe size line. In this example it is shown to be 38" per second.

Pipes should be sized as follows:

Pipe sizes below 2" consult Friction Chart for milinch resistance allowable at or below 48" per second.

Pipe sizes 2" and above maximum of 500 milinch resistance per foot.

Where noise is not a factor, velocities may be higher.

In an area where noise is objectionable follow the maximum allowable velocities as shown

Entrained air in the piping can cause noise at low velocities. Higher velocities can be used providing air is not entrained in the water in the piping.

Fig. 15

For more accurate calculations the piping system should be laid out and the number and type of fittings to be used should be listed. One method of determining a reasonably accurate measurement of "equivalent length" is called the "elbow equivalent" method. In this method the resistance to flow of various pipe fittings is expressed as the ratio of the fitting resistance compared to the resistance of a standard 90° elbow. One elbow of any size is assumed to have 25 times the nominal pipe diameter, feet of resistance head. On this basis a one inch nominal standard elbow would have a resistance equal to 25 inches or 2.1 feet of one inch pipe and a 3 inch standard 90° elbow would be the equivalent of: 25 x 3 = 75 inches or 6.2 feet of 3 inch pipe. Based on the elbow equivalents of various fittings the pipe fitting resistances are expressed in terms of the feet of straight pipe having the same resistance as one of the fittings listed in Table 4.

A supply pipe from a chiller to an air cooling coil is to be 150 feet long and the return is 170 feet long. The flow rate is to be 200 gpm and the pressure loss through the water chiller at that flow rate is 12 feet and the pressure loss through the air cooling coil is 8 feet. Make a preliminary estimate of pipe size and pump selection.

Solution 1

Find pipe size and pipe friction loss. Select a pipe to give 500 mil inches per foot or less. At a flow rate of 200 gpm Fig. 15 shows that a 3 inch pipe will cause a pressure loss of about 950 mil inches per foot. A 3½ inch pipe will cause a loss of just 500 mil inches per foot and a 4 inch pipe would cause a loss of 250 mil inches per foot.

The 3 inch pipe is too small as indicated by the pressure loss rate in excess of a recommended 500 mil inches per foot. The velocity would be about 100 inches per second or slightly above 8 feet per second.

If 3½ inch pipe is available it will be a good selection. If 3½ inch pipe is not available a 4 inch pipe having a loss of about 250 mil inches per foot will be used:

The total length of pipe estimated is:

Supply	150 ft.	
Return	170 ft.	
Total	320 ft.	
50% fitting allowance		160 ft.
equivalent length of pipe		480 ft.

The head loss in an equivalent length of pipe is estimated as:

$$(3\tfrac{1}{2}\text{ inches})\quad \frac{480\text{ ft. X }500\text{ mi per ft.}}{12{,}000\text{ mi per ft.}} = 20\text{ feet}$$

$$(4\text{ inches})\quad \frac{480\text{ ft. X }250\text{ mi per ft.}}{12{,}000\text{ mi per ft.}} = 10\text{ feet}$$

The total head loss based on $3\frac{1}{2}$ inch pipe = 20 ft. piping loss
12 ft. chiller loss
8 ft. coil loss
40 ft.

Based on 4 inch pipe the total head will be 10 ft. less or 30 ft. total.

Table 5A

Impeller Diameter	*Pump Efficiency*	*Flow Rate*	*Head*	*Horsepower*
5"	about 40%	122 GPM	14.5 ft.	less than 1
$5\frac{1}{2}$"	about 57%	139 GPM	20 feet	about 1.5
6"	about 64%	165 GPM	$27\frac{1}{2}$ ft.	about 1.8
$6\frac{1}{2}$"	about 69%	188 GPM	$35\frac{1}{2}$ ft.	about 2.5
7"	about 71%	208 GPM	43 feet	about 3.1

Table 5B

$5\frac{1}{2}$"	58.5%	110	19	1.15
6"	66.5%	187.5	27	1.85
$6\frac{1}{2}$"	70%	212.5	34.5	2.6
7"	73%	237	41	3.4

It is evident that none of the 1510 series pumps whose performance curves given in Figs. 11A, 11B, 11C or 11D, are suitable for 200 gpm at 30 ft or 40 ft of head but a 1531 series pump size 3 inch A.B. turning at 1750 rpm, Fig. 16, will be satisfactory for either pipe selection.

The friction loss through the system is 40 feet using a $3\frac{1}{2}$ inch pipe at a flow of 200 gpm the friction loss at other flow rates being proportional to the square of the flow rates will be as follows:

$$\text{at } 100 \text{ gpm } \left(\frac{100}{200}\right)^2 \times 40 \text{ ft.} = .25 \times 40 \text{ feet} = 10 \text{ feet}$$

$$\text{at } 150 \text{ gpm } \left(\frac{150}{200}\right)^2 \times 40 \text{ ft.} = .562 \times 40 = 22.48 \text{ feet}$$

$$\text{at } 250 \text{ gpm } \left(\frac{250}{200}\right)^2 \times 40 \text{ ft.} = 1.56 \times 40 = 62.4 \text{ feet}$$

A curve plotted on the pump performance sheet is shown as curve "S1", the pressure loss vs flow rate for the system. The intersection of the flow rate curve S1 with the curves showing pump performance at various impeller diameters are shown in Table 5A.

A similar curve "S2" is a system curve based on using 4 inch pipe and shows the performance with the same pump, as in Table 5B.

From the above data it appears that using the 4 inch pipe would allow the use of a 3 horsepower 1531 pump with a 6½ inch diameter impeller and the motor would develop about 2.6 hp and operate with less than the full motor rated current. The flow rate of 212.5 gpm is more than the design rate of flow.

The use of the 3½ inch pipe (if available) would require a 7 inch impeller and operate a 3 hp motor above its rated horsepower although safely within an allowable 15% overload. It would supply about 208 gpm or less water at higher horsepower than if the 4 inch pipe was used.

Example 2

In the first of this series on hydronic cooling a cooling load was calculated for a 100 x 300 x 20 foot building and was found to be approximately 150 tons of cooling. Fig. 17 shows this building as an open light manufacturing space. The machinery room is located on the roof of the building and two horizontal air conditioning cabinets of 75 tons cooling capacity each are hung about 15 feet above the floor and air ducts from these units are used to distribute the conditioned air throughout the building. Two 75 ton package liquid chillers can be piped together on the water sides, or one 150 ton package liquid chiller can be used.

For selecting the piping sizes to be used a flow rate is to be based on 10° temperature rise through the coils at full load. Since the product of gpm per ton x °td. (temperature difference) equals 24, 10° td. indicates a flow rate of 2.4 gpm per ton or 360 gpm for 150

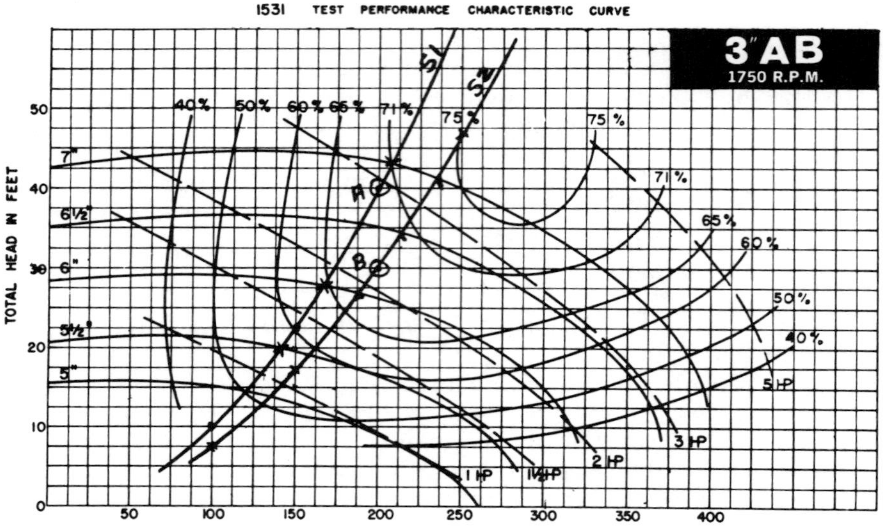

Fig. 16

Table 6

a	*b*	*c*	*d*	*e*	*f*	*g*	*h*
Pipe Run	*Flow Rate*	*Pipe Size*	*Length of Run*	*Elbow Eq.*	*Eq. Length of Run*	*Frict. Rate*	*Total For Run*
	GPM	*Nom ID*	*Feet*	*No. of*	*Feet*	*MI/ff.*	*MI*
A-B	360	5	60	5	112	350	42,000
B-C	180	4	166	.5	170	200	34,000
C-D	180	4	15	1	23	200	4,600
E-F	180	4	15	3	40	200	8,000
F-G	180	4	166	0	166	200	33,200
G-K	360	5	45	7	115	350	40,250
K-A	360	5	25	6	85	350	29,750
			Longest series run total			(i)	191,800
B-H	180	4	15	4	48	200	9,600
J-G	180	4	15	3	40	200	8,000

Table 7

	Supply Lines							*Return*	
Pipe Run	*Flow Rate*	*Pipe Size*	*Run Length*	*Equiv. Elbow*	*Equiv. Length*	*MI/Ft.*	*Total MI/Run*	*Flow Rate*	*Pipe Size*
A-B	360	5	15	1.5	25	350	8,750		
B-C	324	5	60	.5	65	200	13,000	36	2
C-D	288	4	60	.5	62	460	28,500	72	2½
D-E	252	4	60	.5	62	380	23,500	108	3
E-F	216	4	60	.5	62	300	18,600	144	4
F-G	180	4	85	2.5	106	220	23,500	180	4
G-H	144	4	60	.5	65	140	9,100	216	4
H-J	108	3	60	.5	63	310	19,500	252	4
J-K	72	2½	60	.5	63	420	26,400	288	4
K-L	36	2	60	.5	62	300	18,600	324	5
L-W	36	2	10	1.5	16	300	4,800		
W-M	360	5	15	15	82	350	28,500	360	5
M-A	360	5	60	4	100	350	35,000	360	5
						Total	257,750		

tons and the flow rate to each air conditioner coil will be 180 gpm. The steps that are helpful in selecting the pipe sizes and selecting the pump are as follows:

Step 1 — Make a scale plan of the layout. Label points in the piping system. See Fig. 17.

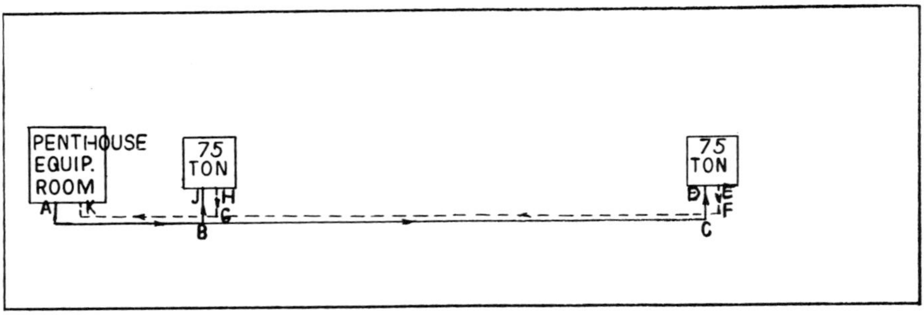

Fig. 17

Step 2 — Prepare a table to list the following information:

a. Piping run designation
b. The flow rate in each run
c. The pipe size selected for each run
d. The length of each pipe run
e. The number of elbow equivalents in each run
f. The equivalent length of each run
g. The friction loss rate in mil-inches per foot
h. The total friction loss in each run of pipe
i. The total friction of the longest series of runs.

Step 3 — Fill in complete table. (see Table 6 obtain or calculate as follows:

a. Fill in pipe run designations from scale layout
b. Flow rates are 360 gpm from A to B and G to K. Others are 180 gpm
c. See pipe sizing chart—select pipe not to exceed 700 mil inches per foot or 8 feet per second velocity at 360 gpm. 5 inch pipe gives 350 mil inches per foot
d. Scale horizontal runs. Estimate drops or riser lengths.
e. From A to B (chiller to first takeoff)

	Length
1. chiller horizontal to drop ell (2. ells)	5 ft.
2. Vertical drop. Pent-house to space below (one ell)	8 ft.
3. Horizontal to main pipe (one ell)	12 ft.
4. Main line to B (one ell)	35 ft.
5 ells +	60 ft.

f. Add equivalent length of fittings from Table 4 to the straight pipe lengths for total equivalent length of run.
g. Pick from pipe friction chart Fig. 15 at time item (c) is established.
h. Multiply equivalent length of run x mil inches per foot for the run.
i. Add (h) for longest series run.

Step 4 – Final piping friction loss in feet of head. Divide item (i) by 12,000 mil inches per foot of head.
Step 5—Find total pump head required at pump flow rate required. Add piping friction head, chiller, and cooling coil friction heads.

Calculation For Steps 4 and 5

Piping friction head (from column H, Table 6) 191,800 Ml per 12,000 Ml per ft. =	16.0 ft.
Estimated chiller head loss	12.3 ft.
Estimated coil head loss	14.3 ft.
Total pump head required at 360 gpm	42.6 ft.

Step 6—Select pump to handle flow rate required at total head required from step 5. Find pump series, impeller diameter, and motor horsepower so that the motor is not overloaded and so that pump efficiency is as high as possible, preferably up to 70% and not below 65%. A system curve can be made up on tracing paper on same scale and ordinates as pump performance curves and placed over the pump performance curves to determine flow head, and efficiency for available impeller diameters. System flow rates can be assumed and head established at each flow rate to plot system curves based on design flow rate and design head.

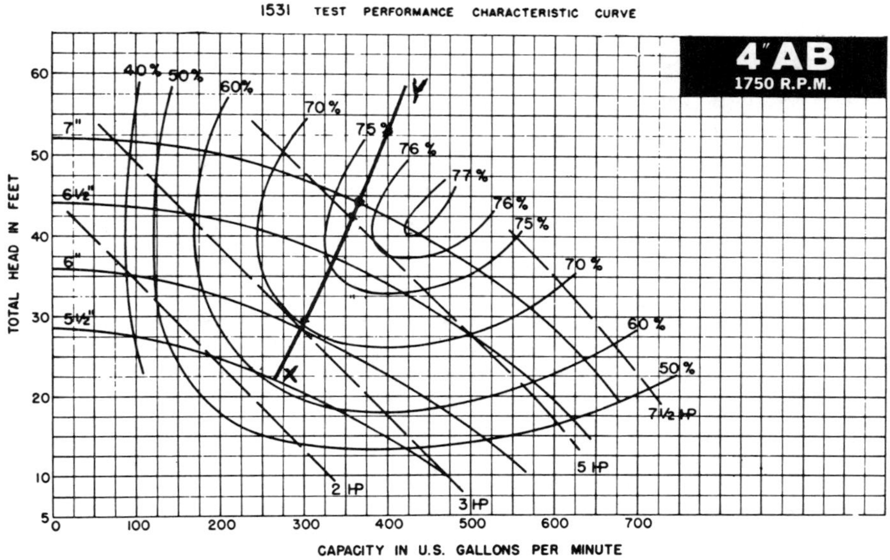

Fig. 18

Head = design head (step 5) x $\left(\frac{\text{gpma}}{\text{gpmd}}\right)^2$ where at any flow (gpma) is assumed flow and (gpmd) design flow.

Refer to Fig. 18. Make up a system curve using 3 points at three flow rates as follows:

Flow Rate	*Head*
300 gpm Head = 42.6 $\left(\frac{300}{360}\right)^2$	= 29.6'
360 gpm (design)	42.6'
300 gpm Head = 42.6 $\left(\frac{400}{360}\right)^2$	= 52.6'

At the intersection of the system flow rate curve "X-Y" and the 7" impeller curve, the flow rate is about 365 gpm, the head is 43.5 feet, the efficiency is between 75 and 76% and the input horsepower will be less than 7½ hp so that the pump selected would use a 7½ hp motor drive.

The pump is a 4" AB, 1750 rpm, 7½ hp, type 1531.

Example 3

Another method of cooling the 100 x 200 foot building would be to use 10 fifteen ton units instead of one or two larger units. In this case the only ductwork required will be for short lengths with directional flow outlets to provide a good air distribution pattern.

Fig. 19 shows 10 suspended 15 ton air conditioners piped in a reverse return pattern so that the pressure loss from the liquid chiller outlet through the supply piping, air conditioning coils and return piping, back to the pump and chiller return, is the same for all chillers. Ten

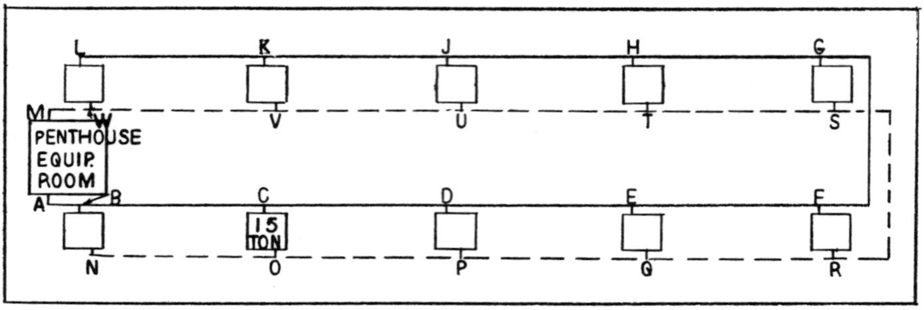

Fig. 19

zone controls can be used to operate modulating face and by-pass dampers or one control may be used to control more than one unit.

Table 4 shows the pipe sizes arrived at, and the flow rates through each supply and return portion of the piping.

Pipe Head 257,750 per 12,000 mi per ft.	= 21.5 ft.
Coil Head Estimated at 10 ft.	= 10.0 ft.
Chiller Head Estimated at 13.2 ft.	= 13.2 ft.
Pump Head	= 44.7 ft.

Reference to Fig. 18 shows that a 7" impeller in a 4AB No. 1531 pump will deliver 350 gpm at a head of about 44.9 feet and will underload a 7½ horsepower motor. It will therefore be a satisfactory pump selection and operate at about 75% efficiency.

Central Station Air Conditioning

Air conditioning has been defined as the process of treating air so as to control simultaneously, its temperature, humidity, cleanliness and distribution to meet the requirements of the conditioned space.

Hydronically, heat can be carried to or from the air being conditioned. When heat is removed from air by means of a heat exchanger being supplied with chilled water the air transfers both sensible and latent heat through the heat exchanger to the chilled water as long as the chilled water is able to keep the heat exchanger surface in contact with the air below the air dew point. If the coil surface is above the dew point but below the dry bulb temperature of the air only sensible heat can be transferred.

For conditions of health and comfort 40 to 60 % relative humidity are considered ideal. Temperatures in the neighborhood of 70 to 75° are excellent for comfort for most people at ideal humidity conditions for summer or winter steady conditions. A person entering a space at 80° from an outside temperature of 95° may feel uncomfortably cool until his body adapts to the 80° temperature after which he may feel too warm. For large stores or other places which customers enter for short periods only, it may be quite difficult to keep the space cool enough for the employees who are in the space continuously without overcooling short time customers who are subject to further shock when they again pass back to the 95° outside temperature.

The temperatures required in a given space may therefore be variable depending upon many factors. The desireable relative humidity depends upon the use of the conditioned space when other considerations than health and comfort are involved. For a printing plant or other applications where static electricity can be very troublesome relative humidities of 60% and over are desirable. For machinery storage or precision manufacture of iron or iron alloy parts, relative humidities of 45% or lower may be required to prevent corrosion. The use of chilled water makes possible the removal of both sensible and latent heat. Warm water may then be used to add sensible heat only with the net result of removing moisture or drying the air.

Air conditioning systems may be classified in several ways. A central system is one in which the air is treated in one central conditioner for distribution by ducts to a comparatively large area. The heat and refrigeration effect may be transferred to water at a remote location and carried to the central air conditioner or conditioners hydronically. There are many combinations of apparatus and control methods used. This article will touch on some of the combinations of equipment and control methods used in central station type air conditioning.

Central Station Systems

In central station air conditioning the air may be conditioned for a complete building or one section or zone of a building in a central apparatus and distributed by means of ducts throughout the building or zone. For example a department store having 10 sales floors each a block square might be divided into forty zones or 4 per floor. Liquid chillers may be located in a second or third basement and develop several thousand tons of refrigeration effect in chilling water which is pumped to the 40 zones throughout the store.

Fig. 20 shows an all year air conditioning cabinet. It is provided with a return air duct which picks up air from various locations in its zone. It has a fresh air supply ducted in from an outside wall or roof top. Supply and return air dampers are interconnected so that the return air damper may be wide open while the fresr air is set at a minimum setting. Movement of the damper motor will operate to close off return air and supply more fresh air.

A cooling coil is provided with chilled water from a liquid chiller as required. The chilled water coil must be supplied with water at a low enough temperature to condense moisture out of the air to provide air at the required dew point as it leaves the air conditioner cabinet.

Central Station Control

If both temperature and humidity are to be controlled in a zone of a building by non-modulating controls for year around operation, heating and cooling thermostats would be required as well as humidistats to control upper and lower humidity limits. If the temperature limits are to be maintained between 70° and 80° and the humidity is to be maintained between 40% and 60% the year around, a simple system of control can be set up to work with a central station air conditioning cabinet as shown by Figs. 20 and 20C.

The fresh air damper has a minimum stop to provide the necessary air for adequate ventilation of the building. As the fresh air temperature

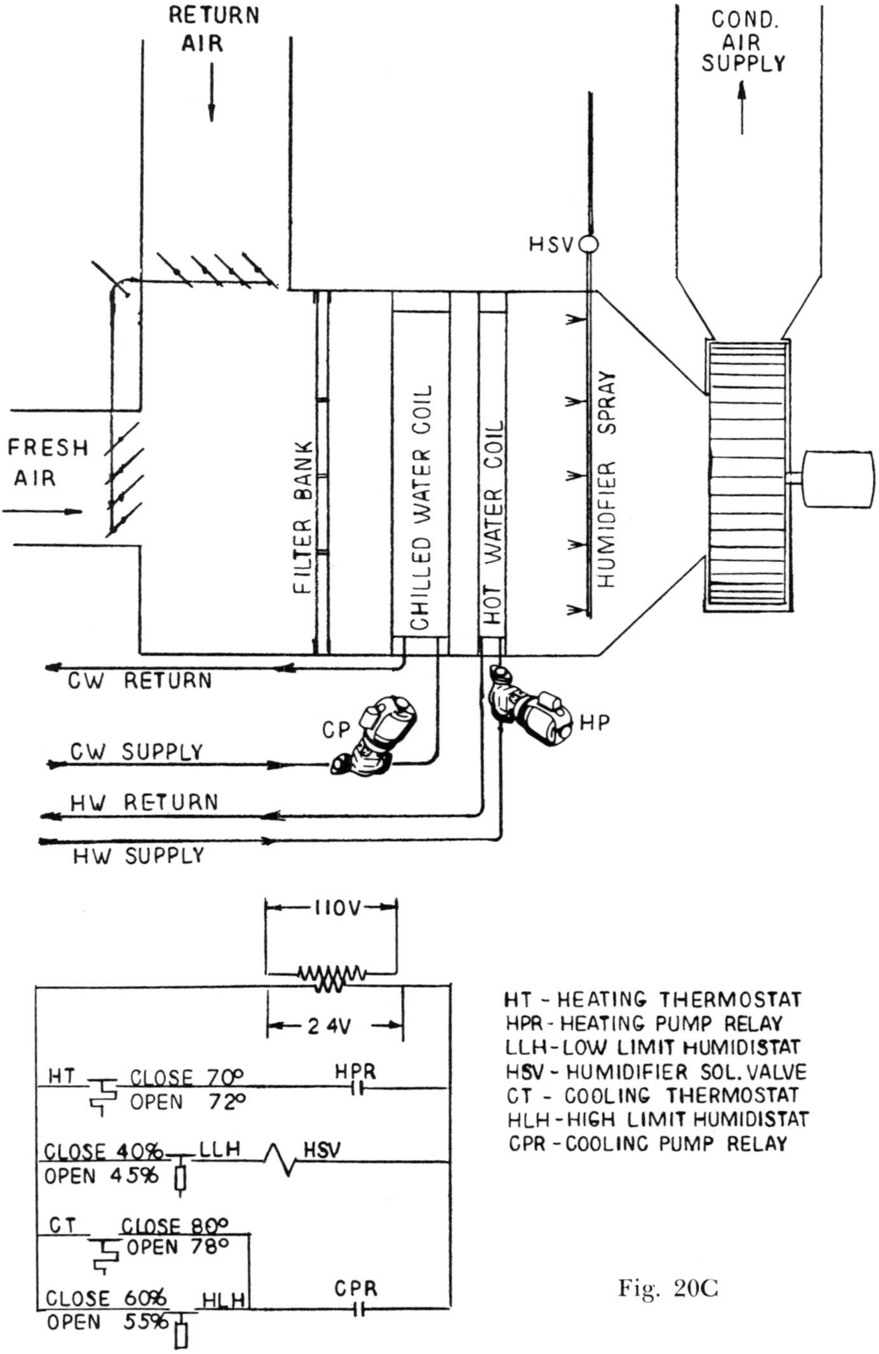
RETURN AIR
COND. AIR SUPPLY
HSV
FRESH AIR
FILTER BANK
CHILLED WATER COIL
HOT WATER COIL
HUMIDFIER SPRAY
CW RETURN
CP
HP
CW SUPPLY
HW RETURN
HW SUPPLY
110V
24V
HT
CLOSE 70°
OPEN 72°
HPR
CLOSE 40%
OPEN 45%
LLH
HSV
CT
CLOSE 80°
OPEN 78°
CLOSE 60%
OPEN 55%
HLH
CPR
HT - HEATING THERMOSTAT
HPR - HEATING PUMP RELAY
LLH - LOW LIMIT HUMIDISTAT
HSV - HUMIDIFIER SOL. VALVE
CT - COOLING THERMOSTAT
HLH - HIGH LIMIT HUMIDISTAT
CPR - COOLING PUMP RELAY

Fig. 20C

exceeds or equals the return air temperature a minimum fresh air setting should obtain and the return air damper will be essentially wide open. The mixed fresh and return air will pass through the filter bank, the chilled water coil, the hot water coil, the humidifier spray section and then pass through the blower and ductwork to the space to be air conditioned.

If the outside temperature is above 80° the return air damper will be wide open and the fresh air damper adjusted to provide the minimum ventilation air required. If at the same time, the relative humidity of the return air is 60%, the high limit humidistat (HLH) will be closed. If the return air temperature is above 80° also, the cooling thermostat (CT) will be closed. With either of these switches closed the zone chilled water pump (CP) will circulate cold water through the chilled water coil. This will cause both sensible and latent heat to be removed from the air.

If the return air temperature drops below 78° the cooling thermostat will open. If the humidity is still above 55% the humidistat will keep the chilled water circulating through the zone coil since more dehumidification is required. In all probability the humidity will drop below 55% before the air return temperature goes below 70° and stop the chilled water circulation. If the return air should drop below 70° before the high limit humidistat opens the heating thermostat (HT) will circulate hot water through the heating coil which will add sensible heat only. With the cooling coil removing sensible and latent heat and the heating coil adding sensible heat only, the relative humidity will be lowered and when less than 55%, will cause the high limit humidistat to open and stop the chilled water pump (CP). If the return air rises to 72° the heating pump will stop so that between 72° and 78°, and between 45% and 55% relative humidity neither the cooling or heating pump will operate.

As the fresh air temperature becomes lower than the return air, the fresh and return air dampers will adjust to provide more fresh air and less return air. In many applications any zones not subject to the influence of outside temperature will require cooling at all times of the year. If the fresh air is below 50 or 60° it may provide the necessary cooling without the use of refrigeration.

With large quantities of cold fresh air and building heat losses the return air temperature will fall below 70° and start the hot water pump to provide heating. The absolute humidity of low temperature outside air (even at 100% relative humidity) will be quite low. For example saturated air at 40° contains only 36 grains of moisture per pound of air and with only sensible heat added to bring it up to 70° where it can hold 110 grains it would have a relative humidity of only about 33%. If the internal moisture load is sufficient to maintain the return air humidity above 40% the low limit humidistat will not close and cause the humidifier solenoid valve (HSV) to open and spray moisture into the air. However if the humidistat (LLH) does close, it will stay closed until the return air relative humidity goes up to 45%. This is in a satisfactory range for health and comfort. At 70° dry bulb temperature and 45% relative humidity the dew point is about 47 to 48° and if the interior surface of a wall or window is below the dewpoint, moisture will condense on it.

The particular control arrangement shown consists of high and low limit thermostats and humidistats all responding to the return air conditions. The damper controls may respond to return air, fresh air, and mixed air temperatures. A provision would also be required to insure against freezing the cooling and heating coils by providing that the mixture of return and fresh air does not become too low. The apparatus and control arrangement as shown in Figs. 20 and 20C essentially may be used to provide controlled temperature, humidity and filtered air for one zone of a building but a number of these zones may be used in one building or on one floor of a building. Several such zones may be provided with chilled water from one high capacity chiller and one source of hot water. Also one such zone may make use of more than one package liquid chiller to provide sufficient chilled water to handle the cooling requirements of a single large zone.

In many cases the relative humidity within an air conditioned zone may be specified to be maintained at or below 45% for all seasons of the year but no lower limit may be required. In such cases the lower limit humidistat (LLH) of Fig. 20C is omitted and the "close" setting of the high limit thermostat will be set down to the high limit permissable such as 45%. This application may be specified for an equipment or storage room to maintain a dry atmosphere to prevent corrosion of iron or its alloys. The humidifier solenoid valve and spray nozzles would be omitted in such applications.

If the conditioned space was to have a dew point of 50° at a lowest temperature of 70°F in the conditioned space the dew point of the air leaving the coil would have to be a few degrees lower so that after the air has picked up moisture in the conditioned space its dewpoint will still be not above 50°F. The air temperature would probably have to leave the chilled water coil at 45° or less and the chilled water would have to be supplied in the neighborhood of 40° or lower.

In most cases supplying air to the room at 45° would tend to reduce the room temperature to less than 70° so the return air heating thermostat will cause the hot water pump to supply hot water to the heating coil, except when the conditioned space has a high sensible heat load such as in extremely warm weather.

In spring and fall the refrigerating equipment may have to operate to remove sensible and latent heat while the heating coil operates to insure against low temperature. During mild winter conditions using a large quantity of outside air and providing heat, may give desired indoor temperature and humidity. For extremely low outside air temperatures, outside fresh air may be mixed with greater quantities of return air and heated to maintain the desired temperature and humidity.

Fig. 21 shows a central station type air conditioning cabinet containing a chilled water coil using face and by-pass interconnected dampers as well as fresh air and return air dampers. The single conditioned air blower supplies conditioned air to more than one zone. Three zones are used in the illustration. Both sensible and latent cooling loads may vary between zones and may shift in magnitude from one zone to another throughout the day. In many instances one zone may require

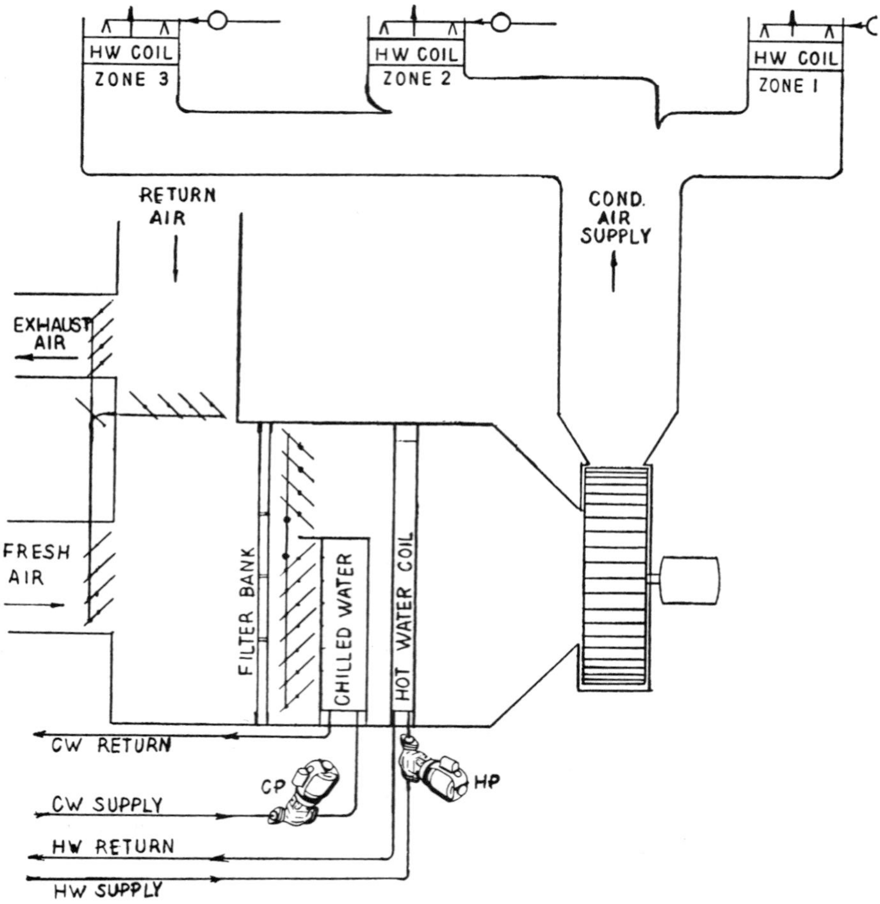

Fig. 21

some cooling at the same time another zone requires heating.

In the system shown, a heating control and low limit thermostat in the individual zone or zone return air duct will serve to insure the temperature will be maintained above the low temperature limit (70°) and the low relative humidity limit (40% or as otherwise desired).

When both air temperature and humidity are higher than desired the chilled water coil face damper will be open and the by pass damper will be closed so that all air passes through the chilled water coil to cause maximum heat transfer. As the return air temperature is reduced a modulating type thermostat causes the face damper to start closing and the by-pass damper to start opening. As the air velocity through the chilled water coil decreases, the latent heat removed increases in proportion to the sensible heat removed. The air through the coil and by-passing the coil thereby tends to have a lower relative humidity after remixing than it would have if none of the air was by-passed. Also the mixed air temperature would be higher than if none of the air bypassed the chilled water coil.

The conditioned air leaving the blower can be maintained at essentially a constant temperature of 60 to 65° during the season when both heating and cooling are required among the 3 zones. If not reheated by the individual zone hot water coils, the air will provide a cooling effect. If one or more of the zone heating thermostats calls for higher temperature it may modulate the water flow or temperature through the hot water coils to produce the desired heating effect in each zone.

Panel Cooling

Panel heating has been done for many years. Commercial and industrial buildings may have warm water coils inbedded in the floor. Wall and ceiling panels may also be used in homes and other applications. However, cooling a home or building by cooling ceiling or wall panels is a comparatively recent application and must take into account special problems.

In cooling seasons air brought into a building may have a dew point between 65 and 70°. Also moisture is added to building air by cooking, burning gas, evaporation from people, etc. and moisture must be removed so that the indoor humidity does not continually increase. If the indoor dew point becomes as high as the lowest window or wall temperature moisture would condense on the walls or windows. Consequently it is necessary for the air to be dehumidified to keep its dew point well below the temperature of any panels used to remove heat from the cooled space.

The air can be dehumidified by means of chemical cartridges or by providing refrigerated surfaces for the moisture to condense on. Such surfaces must be appreciably below the temperature of a ceiling or wall panel used to remove sensible heat from the room.

If the chilled water from a water chiller (40 to 45°) is passed through an air to water heat exchanger in a counter flow arrangement, the dew point of the air in the room may be lowered to as low as 50°F with the outlet water being raised to over 50°. This 50° water may be used in cooling a wall or ceiling panel to 60 to 68° or appreciably above the dewpoint of the air so as to avoid condensation on the panel. With the whole ceiling panel cooled only 10 or 15° below the room an appreciable cooling effect can be obtained.

Air Washers

Dehumidification and air cooling can be accomplished by chilled water in spray cabinets. The air to be cooled, and dehumidified, is passed through a spray cabinet in which it comes into direct contact with a chilled water spray. As the water warms up, the air cools down and the water and air leave at nearly the same temperature.

The air washer also has an air cleaning and odor removing action. The spray water drains into a drain pan and is filtered and pumped back through the package liquid chiller to be rechilled. A system of baffles called an eliminator section is located downstream from the spray section to separate the entrained moisture or mist from the downstream air. Heating is accomplished in a surface type heat exchanger downstream from the eliminator bank.

A comparatively recent development in air washers is a circular spray chamber using an axial flow fan to draw fresh air and return air from

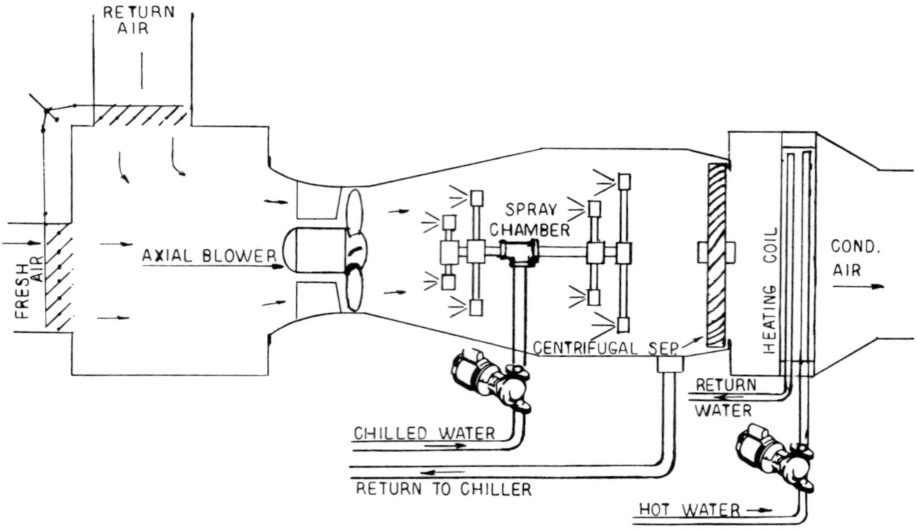

Fig. 22

the mixing box and pass it through the spray chamber in which a bank of rotating sprays (similar to rotary lawn sprinklers) spray the chilled water to accomplish complete coverage of the air passage. (See Fig. 22.)

A rotating eliminator at the air outlet of the spray chamber has a

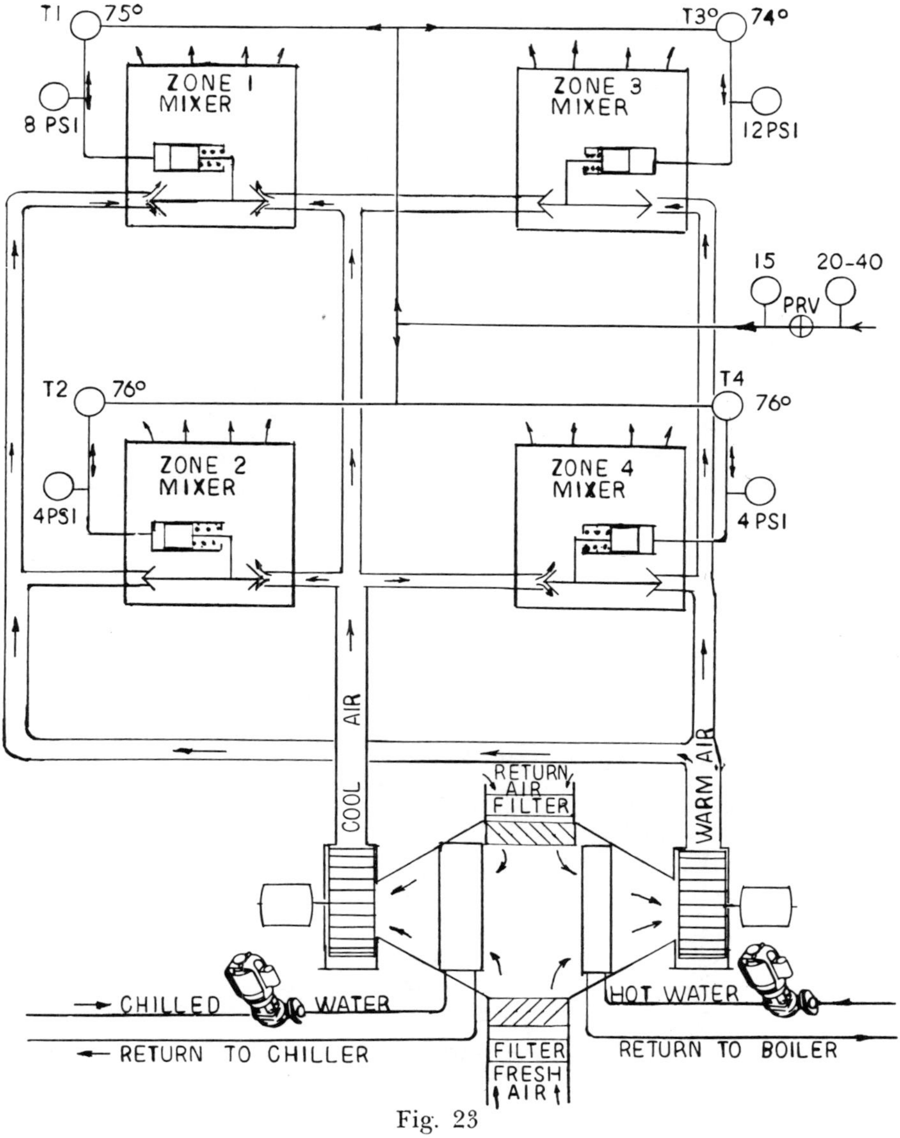

Fig. 23

number of blades which are revolved by the air stream and throw the entrained moisture outward into the casing by centrifugal force. The air velocity through this type washer is much greater than through the rectangular type with vertical sprays and the unit is consequently smaller and lighter.

Fig. 23 shows a two duct system of air conditioning using two blowers. One blower takes a combination of building return air and fresh air and draws it through a chilled water coil. The cooled air is then passed through a duct to several zone air mixers. In the sketch 4 zones are indicated. The second blower draws air from the return and fresh air mixing box through the heating coil and forces it through a duct up to the 4 zone mixers.

FIG. 1A

FIG. 1B

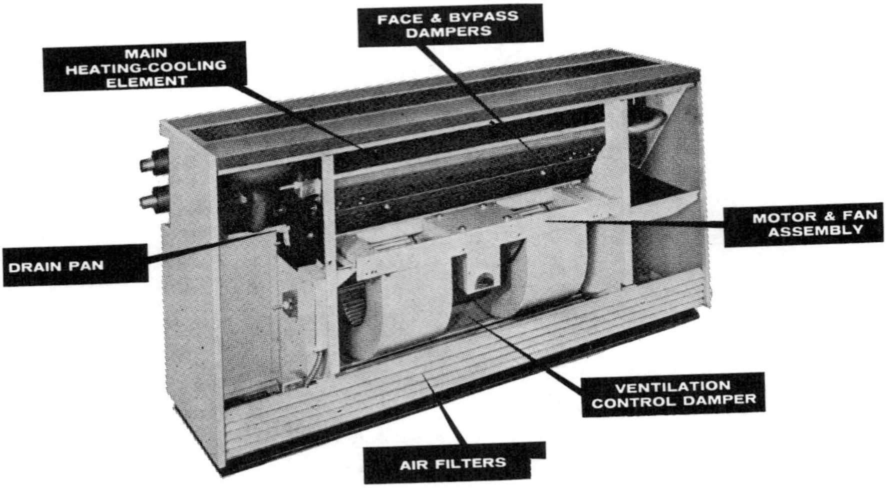

Fig. 25A

The control system illustrated is of the pneumatic type. A small air compressor operated by an electric pressure control is started when the air pressure in a storage tank drops to 20 psi. The compressor then operates to build the air pressure up to 40 psi when the pressure control stops the compressor. The air supply line from the storage tank feeds a control supply at a constant pressure of 15 psi which is held constant by means of a pressure reducing valve (PRV). The supply pressure to each control is constant at 15 psi but the pressure in each control chamber is adjusted to any pressure from 0 to 15 psi depending on the control temperature.

With the set up shown, the control changes the pressure in its control chamber 4 psi for each degree of temperature change. As shown, all 4 controls are set to a 75° setting. At 75° the controlled pressure is shown as 8 psi which holds a spring loaded piston in a cylinder in mid-position so that in zone 1, warm air enters from the warm air duct and mixes with cool air from the cool air duct.

In zone 2, the room temperature is 76° and the thermostat controlled pressure is 4 psi which allows the spring loaded piston to close off the warm air supply duct and provide a maximum opening for the cool air to enter the mixer in an attempt to cool zone 2 from 76° to 75°.

Zone 3 is shown at 74° causing the controlled pressure to be 12 psi so that the cool air valve is closed and the warm air valve is open to provide warm air in an attempt to warm the room up to 75°.

Zone 4 is at 76°, the same as zone 2 so the controlled pressure for thermostat T4 is 4 psi which closes the heating valve and opens the cooling valve fully.

To prevent the unnecessary use of hot or chilled water a thermostat responding to the fresh or outdoor air temperature can be adjusted to shut down the hot water pump whenever the fresh air temperature is above 80 to 85°. Also whenever the fresh air temperature is 60° or lower the chiller water pump (or chiller) may be shut down.

Also the fresh air and return air dampers may be automatically operated to make maximum use of warm outside air to reduce the heating requirements in warm weather and to obtain benefit of cold outdoor air to reduce use of the chiller in cool weather.

Fan Coil Systems

The central station system of air conditioning, conditions the air in a central cabinet and distributes the conditioned air by means of duct work to the air conditioned space or spaces. As opposed to this, the unitary or fan coil air conditioning systems make use of various types of fan coil air conditioning units to condition the air within the conditioned space. In hydronic fan coil systems, heat is added to or removed from air in the conditioned space by heated or chilled water.

Fig. 24A shows a unit heater which includes a heat exchange coil and a fan in a cabinet. This type of heating unit is usually mounted in overhead space and since warm air is lighter than cool air directional vanes may be required to direct the airstream down to the floor or working space. Fig. 24B shows a unit cooler which is essentially the same as a unit heater except that vanes are usually provided to direct the outlet air out horizontally or in an upward path since cool air being heavier, tends to fall. Otherwise the heat exchanger is a unit heater if supplied with hot water and a unit cooler if supplied with chilled water.

Unit heaters or coolers tend to provide low cost heating or cooling equipment. They take up no floor space and are used singly in small spaces or may be used in large numbers in large areas, especially factories, where the fan noise is acceptable. A wall thermostat may be used to stop or start the fan motors while the chilled or heated water flow through the heat exchange coils is continuous.

Unit coolers are provided with insulated drain pans for condensate water which must be piped to drains. The units used only for heating do not require drain pans.

In shipping departments unit heaters are often turned on automatically when the shipping room door opens to provide a curtain of warm air over the door opening in cold weather.

Fan coil cabinets used to heat or cool individual rooms or apartments in hotels, motels, offices, schools, or apartments usually include air filters in addition to surface heat exchangers, blowers, directional vanes, and

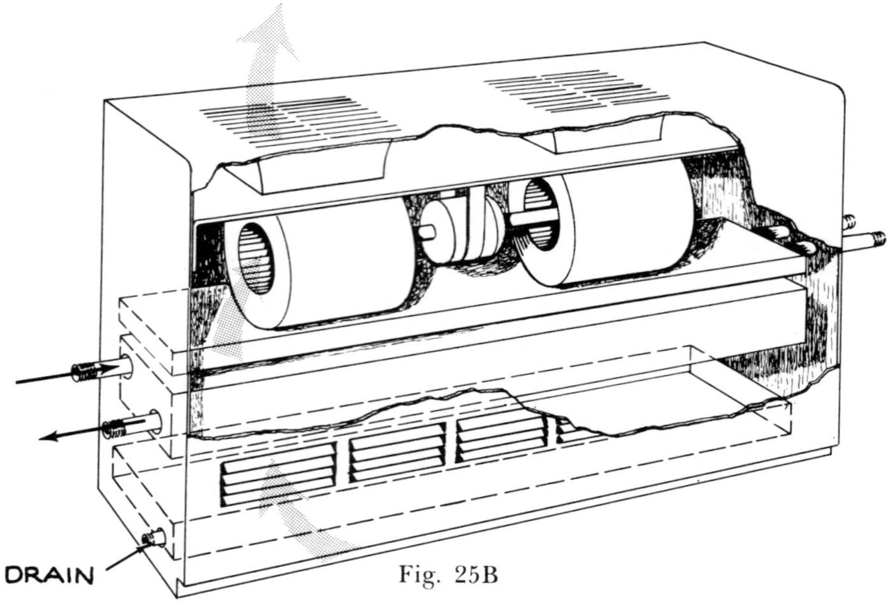

Fig. 25B

dampers. They may or may not provide ventilation requirements. Where heating may be required in one part of a system at the same time that cooling is required in another part separate heating and cooling coils may be provided and a total of four pipes are required to each fan coil to provide supply and return piping for warm water and chilled water.

Figs. 25A, and 25B show fan coil units which are floor, wall, or ceiling mounted. The horizontal ceiling type units may also be built into spaces above hallways, closets, etc. with their controls wall mounted.

Fig. 26 shows a constant flow water circulation through three risers with four floors per riser. The piping is arranged in a reverse return system so that essentially the length of piping from the chiller and return is essentially the same through all coils.

For heating systems the expansion tank is usually tied into the outlet of the hot water boiler along with the proper boiler fittings because the boiler is the highest temperature point in the water system. The point at which air most readily separates from water would be the point of highest temperature and lowest pressure. Consequently, tieing the expansion tank into the suction side of the pump, taking hot water from the boiler provides best air separation.

With hydronic cooling the highest water temperature is in the return line and the lowest pressure is at the top of the return riser and the location shown is ideal for chilled water systems. The return line must be large enough so that the water pressure will increase due to static head more than it decreases due to piping friction.

To obtain a condition in which water would not drop at the desired flow rate due to gravity through a vertical pipe assumes a pressure loss of 12,000 mil-inches per foot or more. If the total vertical and horizontal equivalent lengths of the pipe runs is 24 times the vertical drop in the pipe, the pipe friction could not exceed 500 mil-inches per foot without having a lower pressure at the outlet than at the inlet. In this case pumping energy would be required to force the water downhill at the desired rate.

The pressure reducing valve fed from the "city water" line will feed water into the system whenever the pressure drops below the desired setting assuming that city water pressure is always above that in the system. A check valve in the fill water line is required to prevent city water contamination and loss of system water if the city water pressure should temporarily drop below the system pressure.

The manual air vent is provided at the top of the risers to remove air from the system during the filling operation after which it should be securely closed.

The pressure relief valve is set to relieve excessive system water pressure which can occur especially in a hot water system if the expansion tank should become waterlogged. It is supplied with each liquid chiller unit.

Cooling control may be manual in the case of many motels where such a system is controlled by turning the fan speed control to "off", "low", or "high" settings. If a separate heating system is used such as baseboard radiant or convector hot water system, combination cooling and heating controls may be used to give excellent control. Such a control would be adjustable but could be set for the heating control to close at 72°

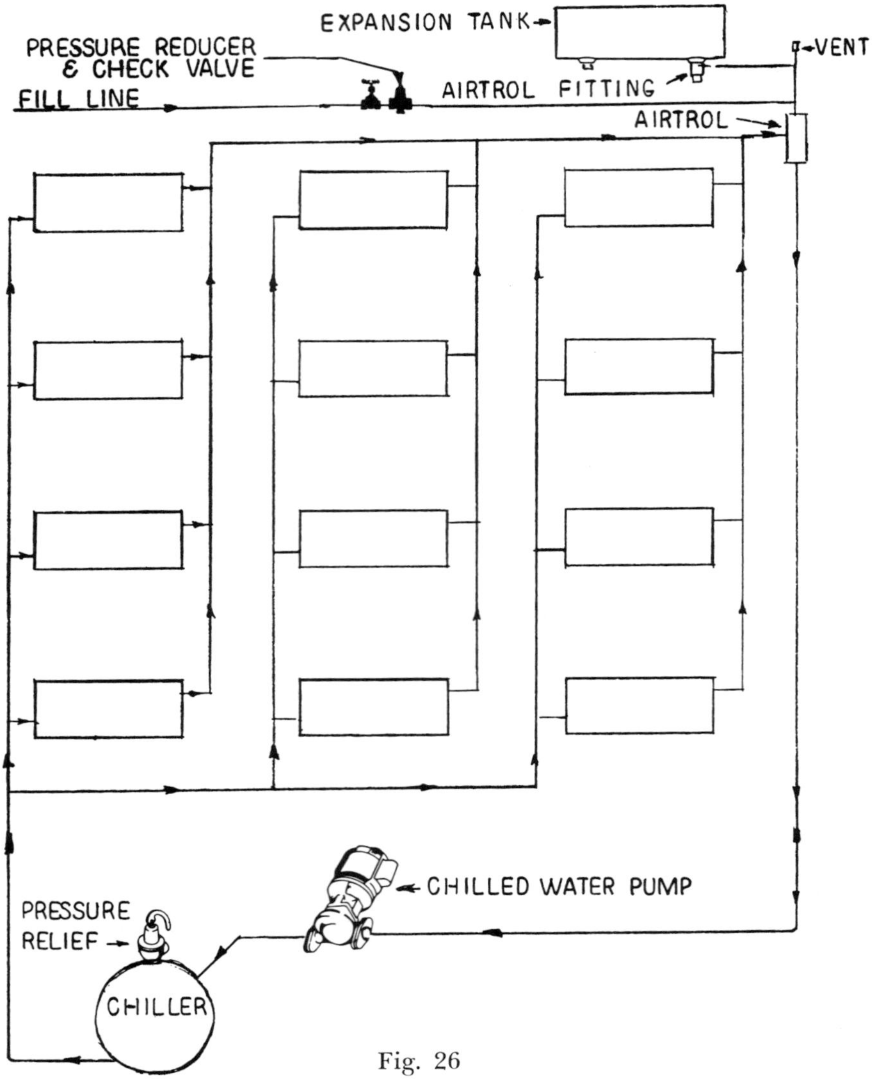

Fig. 26

and open at 74° while the cooling control might open at 76° and close at 78°. In such an arrangement if the room temperature got down to 72° the control would cause a zone control valve to open and circulate hot water through the baseboard heating unit. As the temperature rose above 74° the zone control would stop the flow of hot water. If the temperature rose above 78° the thermostat would cause the blower motors to start forcing air over the chilled water coils. As the temperature dropped to 76° the blower motors would be stopped and neither heating or cooling provided unless the temperature dropped below 72° or rose above 78°.

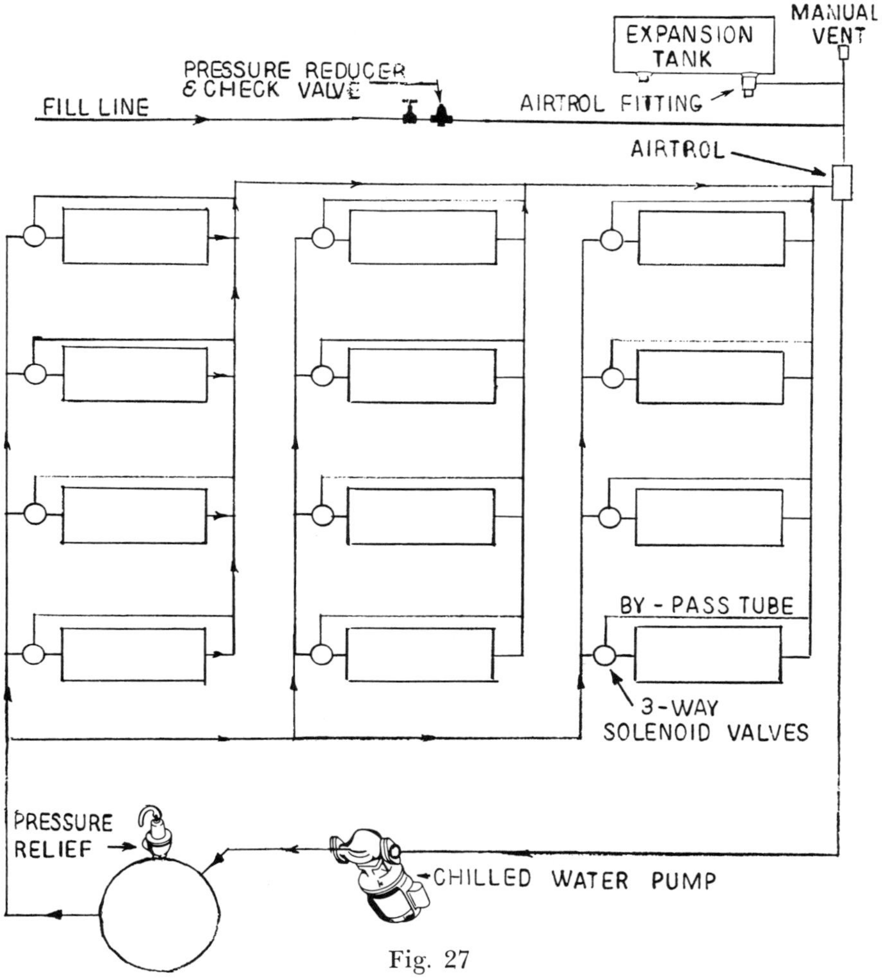

Fig. 27

A constant flow water system such as this can serve any practical number of riser. and number of rooms per riser. For very high buildings it may be wise to limit the number of floors served by one system to a practical limit such that the water pressure is not excessive at the bottom of the risers while not going below about 4 psi at the top of the return risers. For example in an 18 to 20 story hotel or apartment building it may be advisable to use three separate systems limiting each one to six floors.

System Layout

If each system is properly laid out the flow through all cooling coils will tend to be the same and theoretically no balancing valves should be required. However in many cases balancing valves may be used to make adjustments to equalize flow or to provide greater flow in some coils than others to compensate for unforeseen conditions. Low cost square head cocks are often used for this purpose.

Fig. 27 shows a system in which 3-way solenoid valves are used to control the water flow through the air cooling coils. In this system the unit may provide ventilation air and recirculated air constantly and at the room thermostat calls for cooling the solenoid valves are energized and open the passage from the supply riser through the cooling coils and to the return riser to provide cooling and dehumidification. When the room thermostat no longer calls for cooling and is not energized the solenoid valve closes the port to the coil but bypasses direct to the return riser. The by-pass tubes can be selected to have the same pressure drops as the cooling coils at the same flow rates.

Heat or Cool Same Coil

Fig. 28 shows a method of cooling or heating with the same heat exchange surface, and the same hydronic system. Air flow over the coils may be maintained in this system which provides cooling or heating but cannot provide one room with cooling and another with heating at the same time. This system may use an automatic change-over control responsive to outside temperature or master control room to provide the change-over or it may be accomplished by a change-over single pole double throw switch operated by a caretaker.

For an automatic change-over system a cooling setting may be adjusted to activate the coaling system at 78° and deactivate it at 72° while the heating system is activated at 68° and deactivated at 72° with the thermostat acting as a single pole double throw switch so that either heating or cooling is called for.

Individual room cooling and heating thermostats may be set to call for cooling at 78° and until the room temperature drops to 76°. At 72° heating would be called for until the temperature rose to 74°.

The system pump will operate continuously and as long as any zone

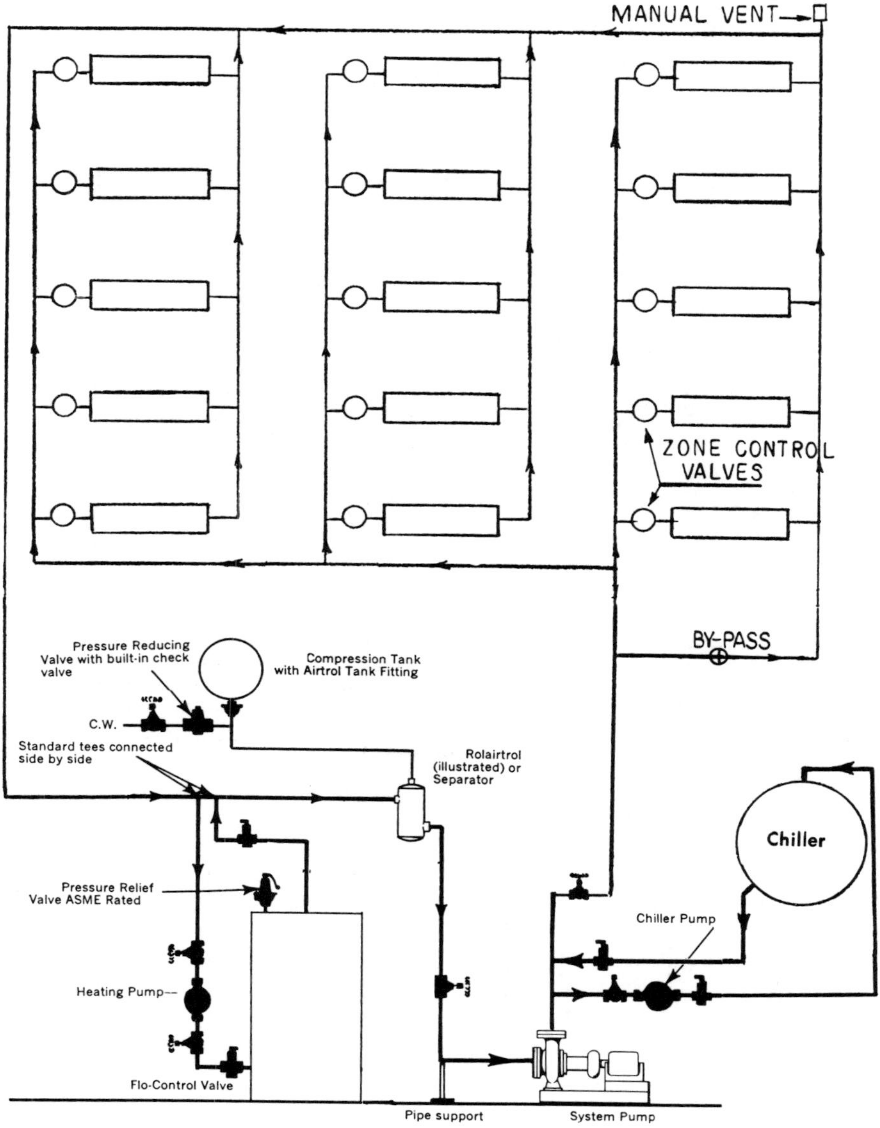
MANUAL VENT
ZONE CONTROL VALVES
BY-PASS
Pressure Reducing Valve with built-in check valve
Compression Tank with Airtrol Tank Fitting
C.W.
Standard tees connected side by side
Rolairtrol (illustrated) or Separator
Chiller
Pressure Relief Valve ASME Rated
Chiller Pump
Heating Pump
Flo-Control Valve
Pipe support
System Pump

Fig. 28

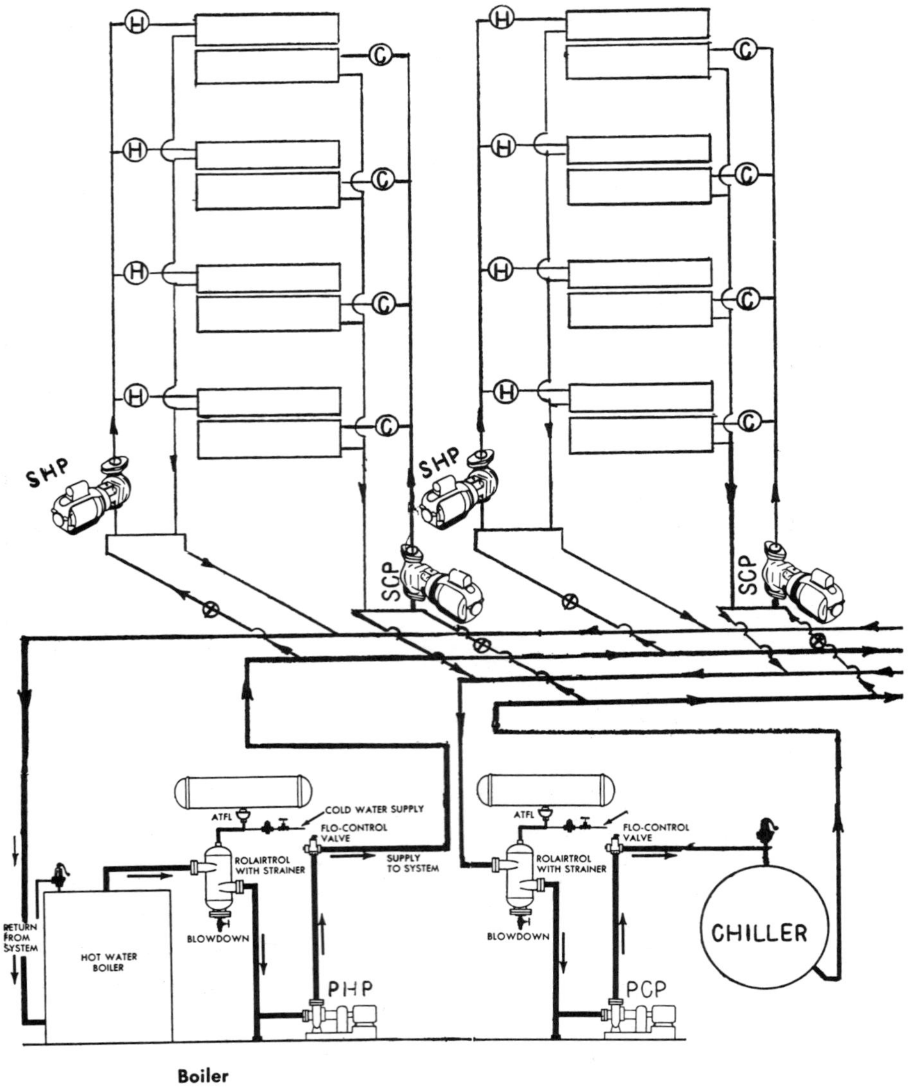
SHP
SHP
SCP
SCP
ATFL
COLD WATER SUPPLY
FLO-CONTROL VALVE
ROLAIRTROL WITH STRAINER
SUPPLY TO SYSTEM
BLOWDOWN
RETURN FROM SYSTEM
HOT WATER BOILER
PHP
ATFL
FLO-CONTROL VALVE
ROLAIRTROL WITH STRAINER
BLOWDOWN
PCP
CHILLER
Boiler

Fig. 29

valve in any room is open, water will be circulated through the pump. If no zone valve is open the by-pass valve would by-pass enough water to prevent pump damage which would occur with continuous operation at no flow.

When the master control responding to a control room or outside temperature calls for cooling the heating system and its primary pump will be shut down and the chiller will operate with its secondary chiller pump to provide chilled water which is circulated through the risers by the system pump and through the chiller by the chiller pump. Also the master control will activate the room thermostats to open the zone valves and provide cooling and dehumidifying in the individual rooms. For any rooms below the temperature at which cooling is required, or in which heating may be called for, the zone control valves remain closed providing neither heating or cooling.

When the master control switches from the cooling to the heating position the chiller and chiller pump will shut down. Also the boiler will be activated to provide hot water and the boiler secondary pump will circulate the water within the boiler system. The system pump will then circulate hot water through the system and the individual room thermostats will open the zone valves on temperature decrease, and close them on temperature increase, if they call for cooling after the chiller is shut down.

Air control equipment is most necessary in connection with heating and only one expansion tank is necessary for both heating and cooling as long as they are tied in together in the same hydronic circuits. This system also uses constant air flow and ventilation for either heating or cooling.

Fig. 29 shows two risers and 4 floors per riser in which heating and cooling coils are provided in each room air conditioner cabinet. As shown, the airflow is from bottom to top of the cabinets with the room air and fresh air (if used) mixed in the bottom of the cabinet and circulated first through the cooling coil and second through the heating coil.

In such a system the heating and cooling units can operate simultaneously. The primary chiller pump (PCP) circulates chilled water through a primary loop and the chiller. Secondary cooling pumps (SCP) are wired in so that as long as any cooling zone control valve on its riser is open the pump will operate and provide cooling water through the coil.

At the same time the primary heating pump (PHP) will circulate heated water through a heating loop and back to the boiler. Any thermostat in a zone calling for heating will cause its riser secondary heating pump (SHP) to operate to supply hot water to any heating zone control valve (H) on its riser which is opened due to its room thermostat calling for heat.

The system thus provides individual room control of either heating or cooling independently of all other rooms. It is also possible to go a step further and by putting high limit humidistats in parallel with the cooling thermostats the cooling coil will cool and dehumidify as long as either control requires. If the humidity control is not satisfied when the heating thermostat calls for heat, both cooling and heating will occur simultaneously. The removal of sensible heat and latent heat by the cooling coil which is partially replaced by sensible heat only in the heating coil, will cause lower relative humidity in the space. If the heating coil was upstream (with respect to air flow) from the cooling coil, no humidity control would be obtained. In this case the heating coil would add sensible heat to the air and the cooling coil would remove sensible heat only.

In review it may be stated that hydronic unitary systems of air conditioning can be worked out to serve a very large number of applications and are ideally suited to buildings which are split up into small rooms or spaces which should be automatically or manually individually controlled.

The minimum function of hydronic unitary equipment is:

(a) Heat and circulate
(b) Cool and circulate

Other functions which can be accomplished by unitary hydronic air conditioning units are:

(c) Dehumidify (Requires drain pan and drain piping)
(d) Filter (Requires provision for filter cartridge)
(e) Ventilate (Requires fresh air opening for outside air)
(f) Control temperature (Individual room controls)
(g) Control high limit humidity (Requires room humidistat and cooling and heating coils
(h) Control low level humidity (Requires humidistat, steam or water supply, solenoid valve and spray nozzle)

Package Chiller Operating Conditions

Thousands of years ago the sun radiated heat to this earth and, with the miracles of life and chemistry, the sun's heat caused a tree to grow, mature, and die and undergo chemical changes until it turned into coal.

A few days ago that coal was removed from the earth and loaded into a railroad car. Today the car of coal was turned upside down and dumped into a hopper and onto a conveyor belt which carried it to a furnace in a power plant. Here the hydrogen and carbon of the coal in the chemistry of combustion, united with the oxygen of the air to form, steam, and carbon dioxide and liberated the heat, chemically stored, in the fuel which came from the sun thousands of years ago.

Heat exchangers called boilers and superheaters, etc. transfer about

85% of the heat stored in the fuel to water, turning it into steam at a high pressure and temperature. Pipes carry the high pressure and temperature steam to a turbine. The turbine turns part of the heat through the science of thermodynamics into mechanical energy. This turns a generator shaft causing most of the mechanical energy to be converted into electrical energy. The heat of the steam, not converted into mechanical energy, was passed to lake or river water by a heat exchanger called a condenser and helps to keep the fish in the neighborhood a little warmer.

In the meantime, getting back to the sun, it radiated heat to the earth a few weeks or months ago also. This heat also caused a plant to grow through the miracle of life and chemistry and this plant also turned into a fuel which we may describe as food. I ate some of this yesterday and today and since my heat engine is operating and my air pump is delivering oxygen to combine with the fuel, some of the sun heat, that was used in growing the food, is being reconverted into heat in my body.

Some of this energy is used to operate my hydraulic machinery to do the small amount of work necessary to make a living, but like all other heat engines I have to dispose of exhaust heat to keep my temperature from going too high and disrupting the rest of the machinery. Assuming this is a hot day in the summer I do this partially by exchanging heat through my heat exchange surface called epidermis from my 99° control temperature to 75° air which surrounds me. I also reject heat to air which I surround temporarily while it is in my lungs. I do this by raising the temperature of the air I breathe and by the evaporation of moisture in my lungs.

If the surrounding air was at 100° I would not be able to reject heat to the air by temperature difference. However my control room would get the signal to turn on the perspiring system so I would reject the required exhaust heat by evaporation. However, the rest of the machinery does not perform so well at high temperature and I would be uncomfortable.

Meantime, back to the sun. That same sun which stored heat to make coal so we could use it to make electrical energy, and that same sun which provided the energy to grow food for me is back. Today it is warming up the outside air. Also it shines on the building I'm in and through the windows in an attempt to make it uncomfortably warm. Since the company I work for wants to keep me comfortable and also because it makes pumps, refrigeration machinery, and heat exchangers it has put some of its products into a mechanical assembly to produce the desired result.

So—the heat which the sun is radiating to this building today, which I'm adding to the building as a result of food I've eaten, (which received its heat energy months ago) plus the heat generated as a result of electrical and mechanical friction, must be removed.

We use conditioned air to dump this heat into, until it gets up to 75 or 80° and then pass the air over a chilled water coil, transferring the heat to the water, warming the water up from 45° to 55° and cooling the air to about 60° to 65°. We pump the 55° water to the "evaporator" or liquid chiller where it dumps heat into refrigerant, boiling at about 37° where the heat is at its lowest (37°) temperature. In compressing the refrigerant the heat temperature is raised and the compressor adds frictional heat also and converts mechanical energy delivered to it into heat and the refrigerant leaves the compressor at the high temperature in the system of 150 to 200°. It soon spills some of this heat (superheat) over into the condenser cooling water and continues to transfer latent heat at approximately 105° to 85° water, warming it up to 95°.

The cooling water at 95° doesn't want the heat either so it is pumped into a cooling tower which sprays it into an air stream which partially evaporates the water. The part of the water which evaporates, picks up its latent heat by reducing the sensible heat of the remaining water and air in the cooling tower until the water is down to 85° again, that is, if the wet bulb temperature is about 75°.

Of course the energy used to run the water chiller was the heat the sun stored thousands of years ago and is being used in the power plant today.

Sometimes we think the song is wrong and that it is actually heat from the sun that makes the "world go 'round". However, getting back to hydronics, the book says that "hydronics is the science of heating and cooling with liquids."

Heating Temperature Differences

In heating with liquids, heat chemically stored in the fuel is usually released, in the process of combustion, at a high temperature and 80 to 85% of it is readily transferred to water in a surface heat exchanger called a boiler. The temperature difference between the hot gases resulting from combustion and the water being heated is very great. If the stack gases were as low as 400° to 600° and the water was heated to 200° the "approach" of the highest water temperature to the lowest hot gas temperature is still a minimum of 200°.

If water at a temperature as low as 150° is returned from a heat exchanger such as a radiator, base board convector or other type of heat exchanger and passes heat to air at 70° to warm the air up to 120°, the minimum temperature difference is 150°-120° = 30°. Thus much more heat exchange surface is required in the radiator than in the boilers. The overall approach of the air being heated up to 120° to the gases leaving the boiler is 400°-120° = 280°, or high enough to effect the required heat transfer readily and high temperature differences allow comparatively low water flow rates to be used.

Cooling Temperature Differences

In contrast, in the cooling phase of air conditioning the heat removed from air being conditioned serves to reduce moisture content of the air as well as reduce temperature. To remove moisture content the heat exchange surface must be below the dew point of the air and frequently the dew point required in air conditioning may be as low as 50° so the entering water temperature must be at 45° or lower to exchange heat with a dew point-to-coil surface difference of less than 5°.

Again comparing heating and cooling, when the heat is obtained by the combustion of a fuel, it runs down hill or is transferred "down temperature" with no expenditure of energy. However, to raise the temperature of heat by refrigeration another source of energy is required to elevate the heat.

In hydronic air conditioning the conditioned air may be cooled to 60° by 35° refrigerant in two stages (air to water and water to refrigerant) so the temperature difference per stage must necessarily be small.

A mechanical refrigeration system picks up heat at a low temperature and rejects heat at a higher temperature and acts to some extent like a liquid pump. The power required to operate a pump varies with the rate of liquid flow and the difference in head through which the liquid is pumped. The power required to drive a refrigeration system varies with the rate at which heat is pumped and the difference in temperature through which it is pumped. Both the initial cost of equipment and the continuing cost of the energy to drive it increases as the difference between evaporating and condensing temperatures increase. Consequently it is necessary to keep heat exchange temperatures as low as possible in order to keep the spread between the evaporating temperatures and the condensing temperatures as low as is economically feasible.

The economics are such that in general, condensers are selected based on a 20° to 30° elevation of condensing temperature above ambient air temperature at design loads. During pull down conditions a 50% increase in absolute suction pressure may cause a normal 30° condenser-to-air-temperature-difference to increase to 45°. With 95° air and 45° higher temperature the condensing temperature is 140° and with R22 head pressures of over 340 psig will occur along with dangerous compressor heating unless oil and/or cylinder cooling or compound compression is used even for evaporation temperatures used in liquid cooling for air conditioning.

Operating conditions at which package liquid chillers operate vary, but within close limits, as compared to commercial refrigeration.

Evaporation Temperature

The temperature difference through which the water, used as a secondary refrigerant, is chilled is usually from 6° to 12° at full load. A fairly standard chilled water supply temperature is 44° with a 52° return temperature. Where humidity control is desired, water temperatures as low as 40° may be required. This is a low practical limit for the chilled water supply.

A good design approach in the evaporator (chilled water supply temperature minus evaporation temperature) is 8° so that with 40° supply the evaporation temperature is 40° − 8° =32°. Lower evaporation temperatures presents the possibility of freezing in part of the evaporator. At operating conditions the evaporation temperature can be quite closely confined between 40° and 32°.

Condensing Temperatures – General

The condensing temperature for a package liquid chiller depends upon the heat exchangers used, and the temperatures of the heat sink into which the heat is to be dumped as well as the medium into which it is to be dumped. The selection of these factors depends upon the availability and climate. Also the ratio of the heat rejected by the condenser to the refrigeration accomplished increases as the compressor pressure ratio increases and may vary from a low of 15,000 Btu per ton to over 25,000 Btu per ton.

City water for Condenser Cooling

Where city water is to be used for condenser cooling, pressure operated water regulating valves are normally used to regulate the flow. When the cooling water is at its maximum temperature and the water cost is low, the water flow may be regulated to allow a 10° temperature rise and with a 10° condenser approach the condensing temperature will be 20° above the supply water temperature at full design load. Where the water cost is comparatively high the water valve may be adjusted to maintain a 20° rise or up to 30° rise which would tend to give condensing temperatures up to 40° above the maximum water temperature. In seasons of lower water temperature the rate of flow will be reduced and the compressor discharge pressure will be maintained reasonably constant.

Cooling Tower Water for Condenser Cooling

Where cooling towers are used they are usually selected for an approach of about 10° which means that the cooling water leaving the tower will be about 10° above the ambient wet bulb temperature. Usually a 10° rise in water temperature through the condenser is used with cooling tower applications and when selecting the condenser for a 10° approach to the condenser leaving water temperature, the combination establishes the condensing temperature as approximately 30° above the ambient wet bulb temperature.

Air Cooled Condensing

Air cooled condensers for air conditioning systems may be used in combination with fans which circulate air so that at design conditions the air warms up 10° in passing through the condenser. The condensing temperature approach to the leaving air temperature is about 20° so that the condensing temperature is established at about 30° above ambient air dry bulb temperature.

Climatic Conditions

Typical dry bulb temperatures, wet bulb temperatures and water temperatures during peak conditions and "normal" condensing temperatures at these conditions can be established for various key cities throughout the country. Assuming an average of 20° temperautre rise where city water is used plus the condensing temperature approach of 10° it can be stated that normal condensing temperature may be:

For city water: 30° + city water + 10°

For cooling tower use: 30° + ambient wet bulb + 10°

For Air Cooled condensing: 30° + ambient dry bulb + 10°

Compression Temperature

High speed reciprocating compressors in large sizes and of present design generally produce more frictional heat than they dispose of due to normal air circulation in the machinery room. The excess of frictional heat generated over that which is passed to the ambient air must therefore be carried away by the refrigerant passing through the compressor. The refrigerant temperature rises in the compressor due to compression heat resulting from compression work and as a result of compressor friction heat being transmitted to it.

Engineers usually think of 300°F as being the maximum safe temperature in a compressor. Temperatures above that have two detrimental effects. (1) The oil break down rate increases rapidly and (2) the yield point of some aluminum alloys decreases rapidly above 300° tending to elongate bearing holes etc.

Many tables are available giving theoretical refrigerant discharge temperatures for the various refrigerants based on saturated suction gas. The curve sheets Figs. 30, 31 and 32 are pressure (psia) vs. entropy plots showing lines of constant temperature in the superheated gas range for the refrigerants R-12, R-22, and R-502 respectively.

By entering the chart from the left at the absolute suction pressure and proceeding to the appropriate suction gas temperature, the entropy of the suction gas is established as indicated on the horizontal axis. Thus, any suction temperature may be used. The theoretical discharge temperature for this entering condition may be determined by proceeding vertically upward to the discharge pressure and estimating the temperature at that point.

Unless some provision is made to remove some heat during compression some of the frictional heat generated in the compressor will be added to the gas causing its discharge temperature to be above the theoretical discharge temperature.

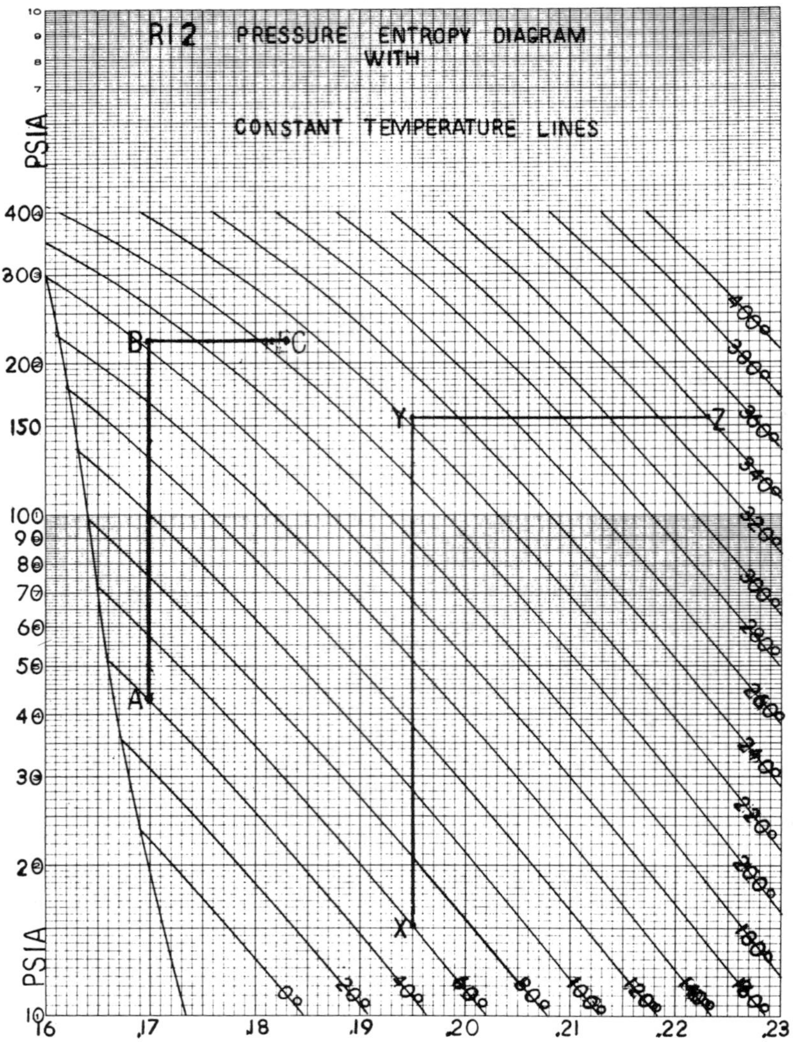

Fig. 30

In an air cooled liquid chiller where ambient air may be up to 110° to 115° dry bulb, the refrigerant may have to condense at 140° to reject heat to the air over the condenser. Fig. 30 shows that with R-12 as refrigerant, (A) 30° evaporation (43 psia suction) and 40° entering the compressor the theoretical discharge temperature (at B) would be 164° at 221 psia (140° condensing). If the compressor was insulated

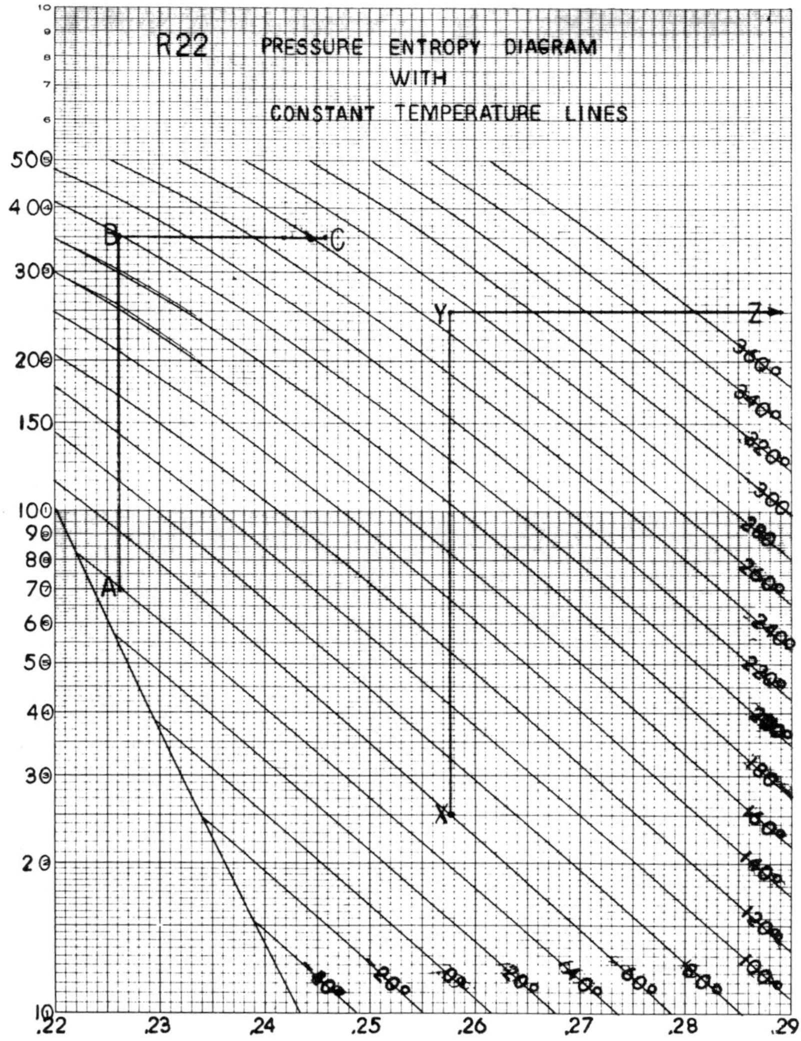

Fig. 31

so that all frictional heat generated in it was passed to the refrigerant being compressed the calculated discharge temperature (B) would be approximately 210°. The temperature rise above the theoretical discharge temperature is read from Fig. 33 and depends essentially on the pressure ratio of the absolute discharge pressure to the absolute suction and upon the refrigerant characteristics.

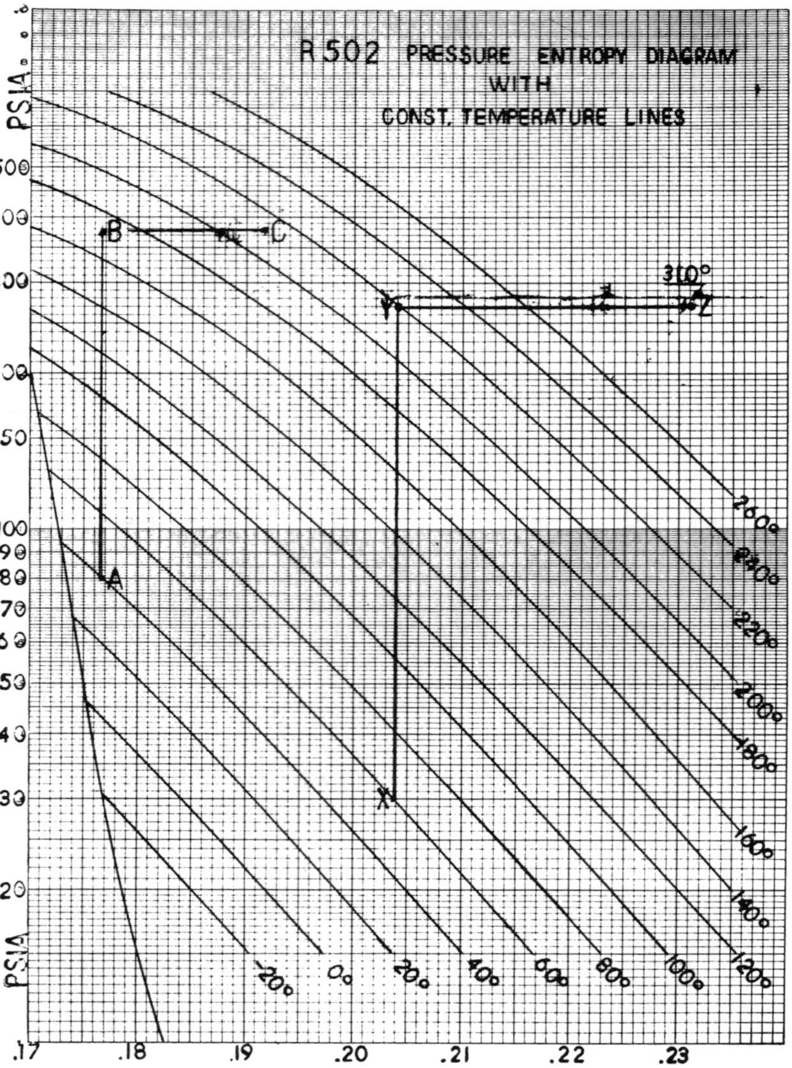

Fig. 32

For the three refrigerants noted all based on 30° evaporation, 40° entering the compressor as at "A" in each of Figs. 30, 31 and 32 and all based on 140° condensing the points "B" show theoretical discharge temperatures with no heat transferred to or rejected by the refrigerant during compression. Points "C" show the final compression temperatures if all compressor frictional heat is transferred to the refrigerant.

If part of the frictional heat is passed to ambient air the discharge temperature will be between that at "B" and "C".

It is noted from the above that the discharge temperature is in the "safe" range (below 300°) with R-22 which has the highest discharge temperature at the most unfavourable condition. (100% compressor frictional heat added to refrigerant). However, if the suction gas was raised in temperature from 40° to 80° due to a long suction line in a warm room the discharge temperature would also go up about 40° to about 305° at the most unfavourable condition.

The points X, Y, and Z on the curve sheets show comparable conditions for —20° evaporation, 60° suction gas, and 112° condensing for all three refrigerants. At these operating conditions an insulated compressor would have discharge temperatures approximating 310° with R-502, 340° with R-12, and above 400° with R-22.

The R-22 system would be very likely to develop trouble in a short time due to high temperature, with R-12 and R-502 operating for longer periods before developing trouble due to excess temperature.

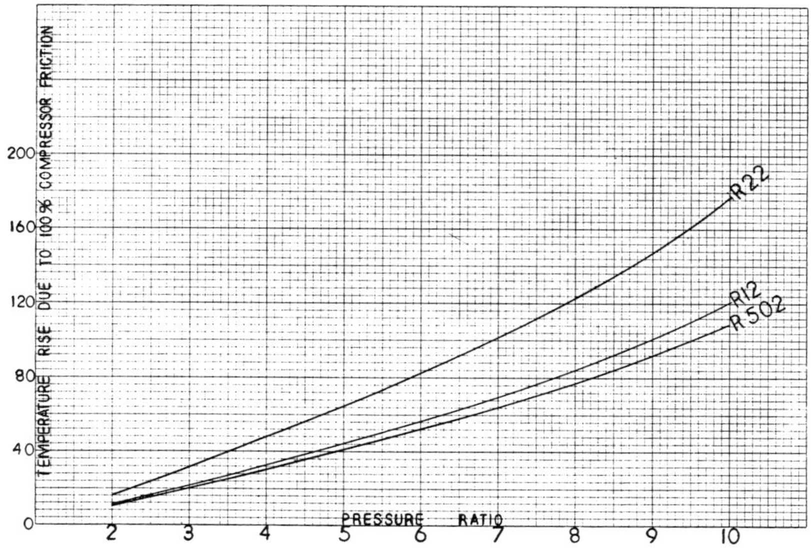

Fig. 33

Where pressure ratios exceed 8 to 10, an oil cooler may be used to remove frictional heat from the compressor lubricant to reduce the temperature difference between "Y" and "Z". Also the judicious use of refrigerant injection can be used to reduce the suction gas superheat.

To review package liquid chiller operating conditions at full load, the following conditions may be considered a guide.

1. Water temperature reduction in chiller 6° to 12°.
2. Water temperature reduction in cooling tower or rise in condenser 8° to 12°.
3. Temperature of water leaving tower 8° to 10° above ambient wet bulb temperature.
4. Evaporating temperature 6° to 10° below chilled water supply temperature.
5. Condensing temperature about 30° above ambient wet bulb temperature if cooling tower is used.
6. Condensing temperature about 30° above ambient dry bulb temperature for air cooled condenser.
7. The compressor discharge temperature for R-12 or R-502 will normally be from 25° to 60° + S.S.H. + condensing temperature and for R-22 it should be from 60° to 120° + S.S.H. + condensing temperature where S.S.H. is the degrees of superheat in the suction gas entering the compressor.

Liquid Chiller Components

A package liquid chiller consists principally of two heat exchangers and a compressor plus the necessary power generator to drive the compressor. The two types of mechanical power commonly used are electric motors and natural gas engines. In addition to these main components there are several refrigerant and electrical controls used in operation and protection of the equipment.

Fig. 34 shows the main components of a package liquid chiller making use of a water cooled condenser. The compressor takes low pressure vapor from the evaporator, compresses and delivers it to the condenser shell where it condenses into a liquid and passes through a fiter-drier, a solenoid valve, and a thermostatic expansion valve into the evaporator tubes.

The refrigerant circuit is very short. The suction line is less than 10 feet long of comparatively large diameter tubing and the pressure loss in the lines is negligible. The short lines, however, introduce the problem of preventing liquid slugs from carrying over from the evaporator to the compressor. This problem is minimized by the use of a separating device in the suction line which tends to separate liquid refrigerant and oil from the suction gases. The liquid is then fed back into the suction line at a restricted rate which the compressor can

Table 8 – Tabulated Conditions for Three Refrigerants

	R-12	R-22	R-502
Evaporation temp.	30F	30F	30F
Evaporation Pressure	43 psia	70 psia	80 psia
Temp. into Compressor (A)	40F	40F	40F
Condensing Temp.	140F	140F	140F
Condensing Pressure	221 psia	350 psia	374 psia
Theortical Discharge °F (B)	164F	200F	172F
Pressure Ratio	5.15 to 1	5 to 1	4.68 to 1
Temp. rise due to 100% Compressor Frict. heat	46°	65°	40°
Temp. with 100% Compressor heat transferred to refrigerant (C)	210°	265°	212°

tolerate without damage. A solenoid valve in the restricted liquid return line prevents the liquid return when the compressor is not operating.

Also the electrical control system provides a "pump down" cycle in which the temperature control closes a liquid line solenoid valve when it no longer calls for cooling rather than to stop the compressor directly. As the compressor pumps the liquid out of the evaporator the lowering of the suction pressure causes the compressor to shut down after the liquid quantity in the evaporator is minimal. This tends to prevent liquid slugging at startup.

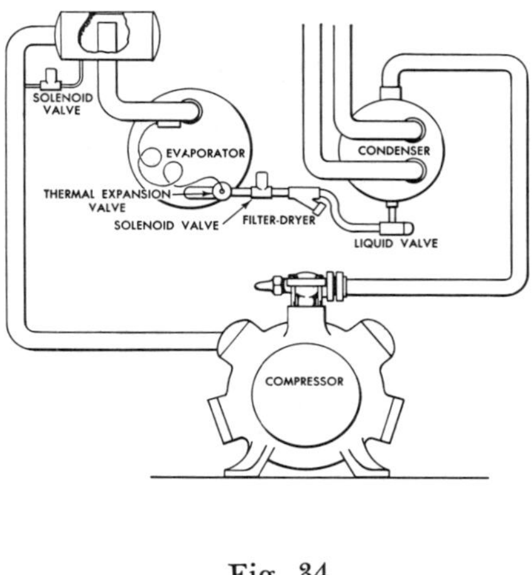

Fig. 34

Compressor

The compressor is the heart of the refrigeration system. It consists essentially of pistons which move up and down in cylinders. A suction check valve admits gas into the cylinder from the suction line and prevents the gas from returning to the suction side of the compressor. A discharge check valve allows the gas to escape to the compressor discharge after compression and prevents the return flow from the compressor discharge back into the compressor cylinder. From 2 to 8 cylinders commonly operate in parallel to provide the desired gas pumping capacity.

The capacity of a refrigeration system depends upon the product of the rate of gas flow in terms of pounds per minute and the refrigerating effect per pound. Since the capacity required by the refrigerating system is highly variable it is desirable to have a variable rate of gas flow. The capacity rate can be decreased most readily by decreasing the number of compression strokes per minute. This can be done by varying the compressor shaft speed or by unloading part of the compressor cylinders or both.

Most of the electric motors used to drive the compressors on package liquid chillers are three phase induction type and are essentially constant speed at all loads. Variable speed can be obtained in special 3-phase motors using wound rotors, slip rings, and external resistances but the cost and bulk of such equipment makes it impractical for this purpose. Motors may also be built for two speed operation but require expensive and bulky switching equipment to change speeds and are seldom used. Variable speeds can be obtained in belt driven equipment also but where 3-phase a-c power is used to drive compressor motors the lowest cost method of varying the number of compression strokes per minute is to unload cylinders.

Where gas engine drives are used the speed can be governed by the engine so that the compressor capacity can be modulated from 100% down to about 50% by engine speed reduction and further capacity reduction is obtained by clinder unloading. The most common method of cylinder unloading is to hold the suction check valve open so that the gas from the suction side of the compressor is drawn into the cylinder on the suction stroke but is expelled back into the suction side through the open valve on the compression stroke.

If the same gas continually passed in and out of an unloaded cylinder it would accumulate frictional heat and become very hot in a short time so that it is necessary that at least one cylinder remains loaded to draw fresh gas through the compressor to carry the frictional heat away.

Fig. 35 shows a cut-a-way view of a typical 4 cylinder compressor. It has a two throw crankshaft each throw accommodating two connecting rods (a) the drive end of the shaft has a carbon against steel shaft seal assembly with the carbon ring sealed to the shaft (b) and the steel nose plate gasketed to the seal end bearing housing (c). The opposite end of the shaft drives the oil pump (d). This pump supplies oil to the main bearings, the shaft seal, connecting rods, and up to the wrist pins.

The cylinder sleeves (e) fit into the counterbores in the crankcase casting (f) and are sealed to it by neoprene O-rings (g). The overload springs (h) hold the sleeves in place but in case of a liquid slug may allow the whole sleeve assembly to move upward to prevent piston or valve plate breakage.

A crankcase heater is shown with its leads (i) near the bottom of the crankcase. This minimizes refrigerant absorption by the oil during shut down periods to prevent foaming and oil slugging at start up. The valve (j) can be used to drain oil out of the crankcase or used as a filtered oil return where an external oil filter may be used. A plug (k) may be provided with a magnetized extension to remove magnetic particles from the oil.

Fig. 36 shows a section of one cylinder of a compressor. The "cylinder" (1) is a sleeve sealed to the crankcase by an O-ring (2) it is held in place by the heavy spring acting between the cylinder head (3) and the discharge valve assembly (4) this allows the sleeve to move if a heavy slug of oil or refrigerant liquid occurs. The unloader assembly consisting of the housing (5), a cylinder (6) and the piston (7) is

Fig. 35

secured to the cylinder sleeve before it is inserted into the crankcase casting. The upper rim of the unloader piston bears on the push ring (8) and as shown, the piston, push ring, and unloader pins (9) are holding the suction ring valve (10) off the seats (11). This results from the force supplied by unloader springs (12). The suction gas enters the compressor cylinder, when the piston (13) is on the downstroke, through the annular space (14) and suction manifold surrounding the cylinder sleeve. On the upstroke the gas reverses its flow through the same passages so that it is not compressed and discharged through the discharge valve consisting of the ring valve (15) and the valve seats (16). When the cylinder is to be "loaded," hydraulic pressure from the

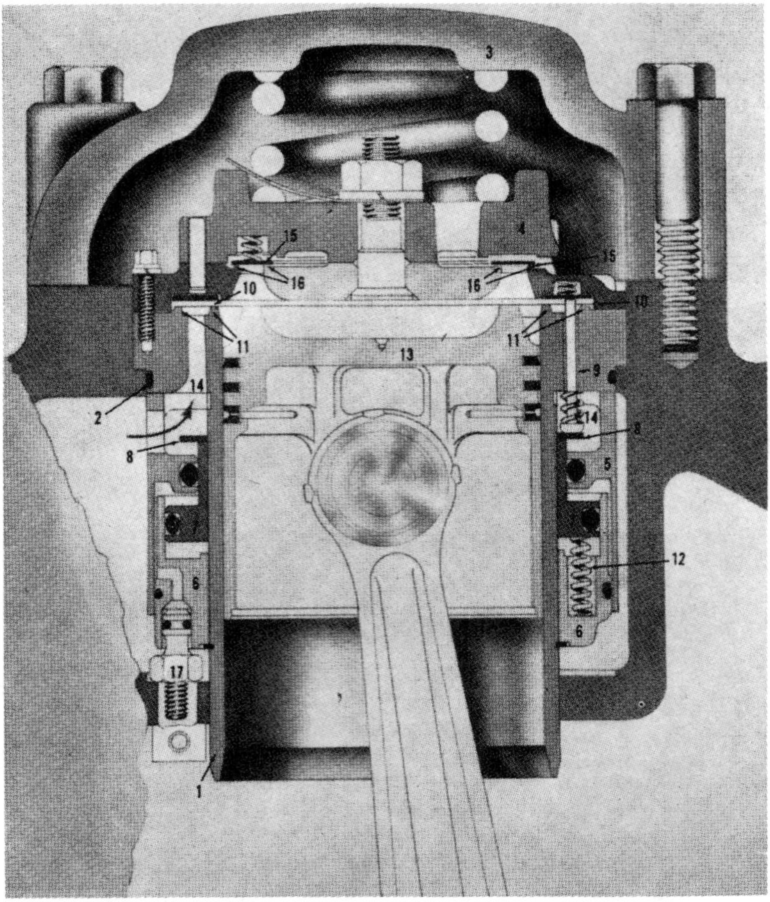

Fig. 36

unloader manifold shown in Fig. 37, is carried through a tube in the crankcase to the unloader nozzle (17). The oil passes between the unloader housing (5) and the unloader cylinder (6) to the space above the unloader piston (7) which is sealed to the unloader housing and piston by O-rings. The piston having the oil pressure on top and only crankcase pressure below is forced downward against the unloader springs (12) so that the push ring (8) and unloader pins (9) follow leaving the suction ring valve to seat normally, thereby loading the cylinder.

Fig. 37 is a schematic illustration showing how the solenoid valves feed oil pressure into the unloader tubes to load the cylinders B, C, and D or relieve the oil pressure to the unloader tubes to unload them. Cylinder A loads independently of all solenoid valves as soon as compressor oil pressure builds up. This allows all cylinders to start unloaded but assures that at least 25 percent of the cylinders load after the compressor oil pump causes a pressure build up.

A spring loaded pressure relief valve (not shown) is secured into the compressor which will by-pass discharge gas back into the suction side if an excessive pressure difference occurs. This is a safety device to prevent damage which might result if the compressor discharge service valve is left front-seated when the compressor is started.

A high capacity cylindrical strainer (not shown) is located under the suction service valve to trap foreign particals which might otherwise get into the valves or bearings.

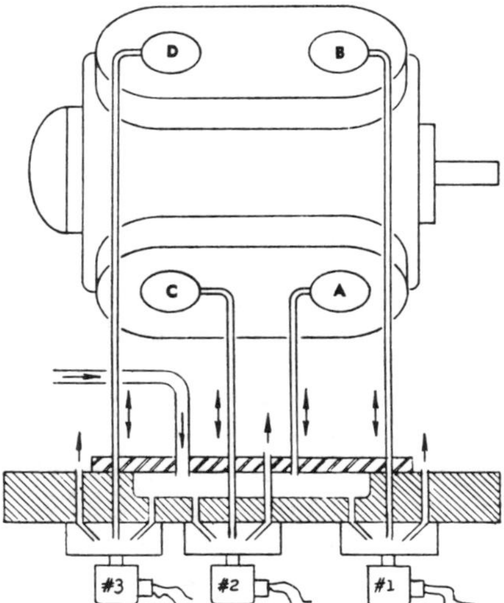

Fig. 37

Fig. 38

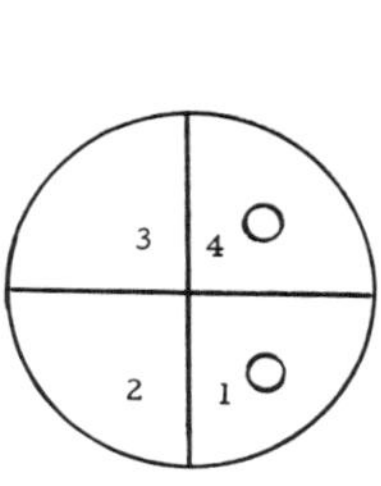

Fig. 38A

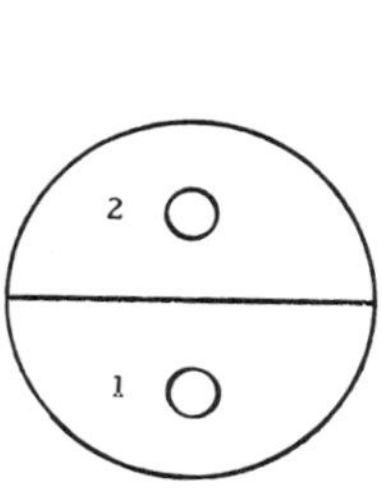

Fig. 38B

Condenser

The water cooled condensers used in combination with cooling towers are of the shell and water tube types, Fig. 38. The cut away condenser shown is arranged for four water passes from end to end through the tubes. In Fig. 38A the four passes are numbered. The inlet water from the cooling tower enters the condenser head and passes from the inlet through all the water tubes entering the tube sheet in quadrant number 1. The water passes to the rear of these tubes and is guided by the rear baffles to the tubes in number 2 quadrant. The water comes forward in the tubes in number 2 quadrant and is guided by baffles to the tubes in number 3 quadrant which carries it to the rear. The rear baffles direct it to the tubes in number 4 quadrant from which it passes forward to the outlet and back to the cooling tower.

With the four pass arrangement all cooling water is passed through a series of four tube banks and each tube bank consists of 1/4 of the total number of tubes in parallel.

Where this arrangement causes too much pressure drop in the cooling water a two pass condenser (Fig. 38B) is accomplished by a change in heads and gaskets. This provides twice as many tubes in parallel and reduces the series length to twice that of a single tube. The two pass condenser also cuts the water velocity in half and the series tube length in half assuming the same flow rate and number of tubes as in the four pass condenser.

The heat transfer from the tube interior to water is very high and would not be materially improved by internal fins. However, the heat transfer from the condensing vapor is lower than for all liquid so that an improvement in heat transfer is effected by the use of a shallow spiral fin on the condenser tube exteriors.

Since the cooling water passes through the tubes, they can become fouled up by water impurities and chemical reactions. Consequently they are constructed so that both front and rear heads can be removed so that interiors can be mechanically cleaned without affecting the refrigerant contained in the shell.

In shell and tube condensers, the lower part of the shell also serves as the liquid receiver. One of the most reliable methods of determining whether or not the refrigerant system is adequately charged, is to determine if a liquid level exists in the receiver. Also there can be some swirling take place so that with an inadequate liquid charge gas can enter the liquid line along with liquid to provide erratic expansion valve action. With the condenser shell shown, the liquid level is adequate if maintained between the two lower test cocks in the shell. If the lower test cock emits liquid when cracked open while the upper of the two emits gas only, the proper operating charge is assured.

There is no particular problem in refrigerant distribution in a shell and tube condenser because as refrigerant vapor condenses on the tube other vapor flows in to fill the space left. There is an advantage in good drainage of condensed liquid. If the drainage of condensate is poor a greater amount of refrigerant will cling to the tubes requiring a greater charge to insure 100% liquid to the shell outlet.

Evaporator

Fig. 39 shows one evaporator design. This design makes use of a single tube sheet and the hairpin shaped tubes have an entrance in the lower part of the tube sheet while the outlet end of each tube is secured into an opening in the upper part of the tube sheet. A baffle in the evaporator head separates the inlet and outlet sections. The refrigerant enters the tubes as a liquid and vapor mixture from the expansion valve. By weight the mixture is usually 20 to 35% gas and 65 to 80% liquid. However, by volume it is over 95% vapor and less than 5% liquid so that it is essentially a wet mist similar to that from a compressed air paint spray gun. Its volume increases at the outlet of the evaporator to about four times that at the inlet. In the evaporator shown the velocity through the tubes increases progressively as the percent of vapor increases. Heat transfer from the tube interior to the mist passing through is comparatively slow as compared with the heat transfer from the water in the shell to the tube exteriors. Consequently the interior heat transfer is increased by the use of internal fins which put more surface in contact with the evaporating "mist." (Fig. 40)

Other designs of "dry expansion" evaporators use straight through tubes and head baffling arrangements tending to increase the number of

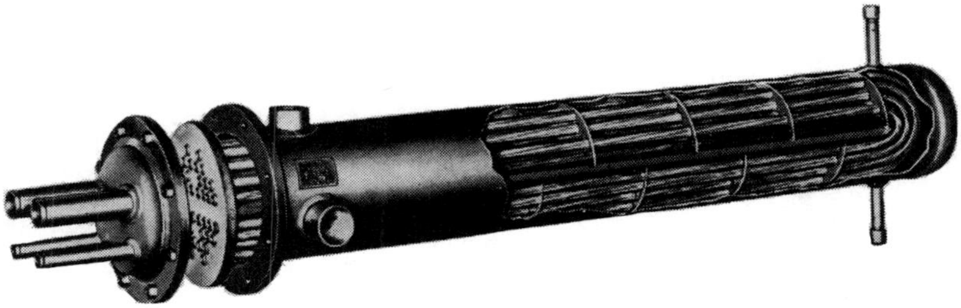

Fig. 39

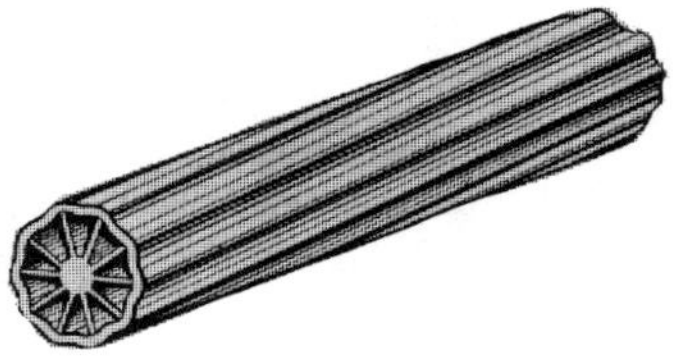

Fig. 40

tubes in parallel in succeeding passes in relation to the increasing gas volume thereby maintaining more nearby constant velocity through the tube passes.

Still others use serpentine tube arrangements and refrigerant distributors so that several tube lengths may be used in each of a number of parallel circuits.

In each of the dry expansion systems the refrigerant evaporates in the tubes while the water to be chilled is guided by baffles to make several passes back and forth across the tubes. The back and forth passes are for one half of the shell and above a center baffle. An equal number of passes are made below a center baffle.

The largest sizes of liquid chillers use centrifugal compressors and are assembled at the site where they are to be used and are usually of a flooded type in which the water to be chilled passes through a bank of tubes while the refrigerant evaporates in the shell around the tubes. This type of chiller has the advantage of a very high heat transfer rate, about double that of the dry expansion chillers. They also have the disadvantage of requiring a much greater refrigerant charge.

High side float valves are used instead of thermostatic expansion valves for reducing refrigerant pressure from the condenser to the evaporator and suction lines may be short and appear as passes in connecting castings rather than as tubes or pipes.

If the refrigerants used, have evaporating temperatures at atmospheric pressure well below 32°, a bad leak to atmosphere can cause water freeze up in the water tubes although the compressor may not be running.

Motors

The two increment winding motors used to power the compressors are not unusual in the 25 to 100 horsepower range. Two complete windings are provided and two motor starters or contactors are used. With the compressors starting without load the motor will come up to speed when the first increment winding is energized. After one to three seconds time delay the second increment winding will be energized and the motor is fully on the line before the compressor loads. The initial inrush current is reduced in this manner.

Controls

Controls for electric package liquid chillers vary slightly between the various manufacturers of such equipment. However, the controls described here are essentially typical. An operating control panel is shown in Fig. 41. The purpose for each of these controls is as follows:

(a) Low Pressure Control. This control shuts down the motor, when it opens, due to low suction pressure. It has a reasonably close differential and normally will reclose within a short interval. However, a non-recycling circuit prevents the motor from starting again unless

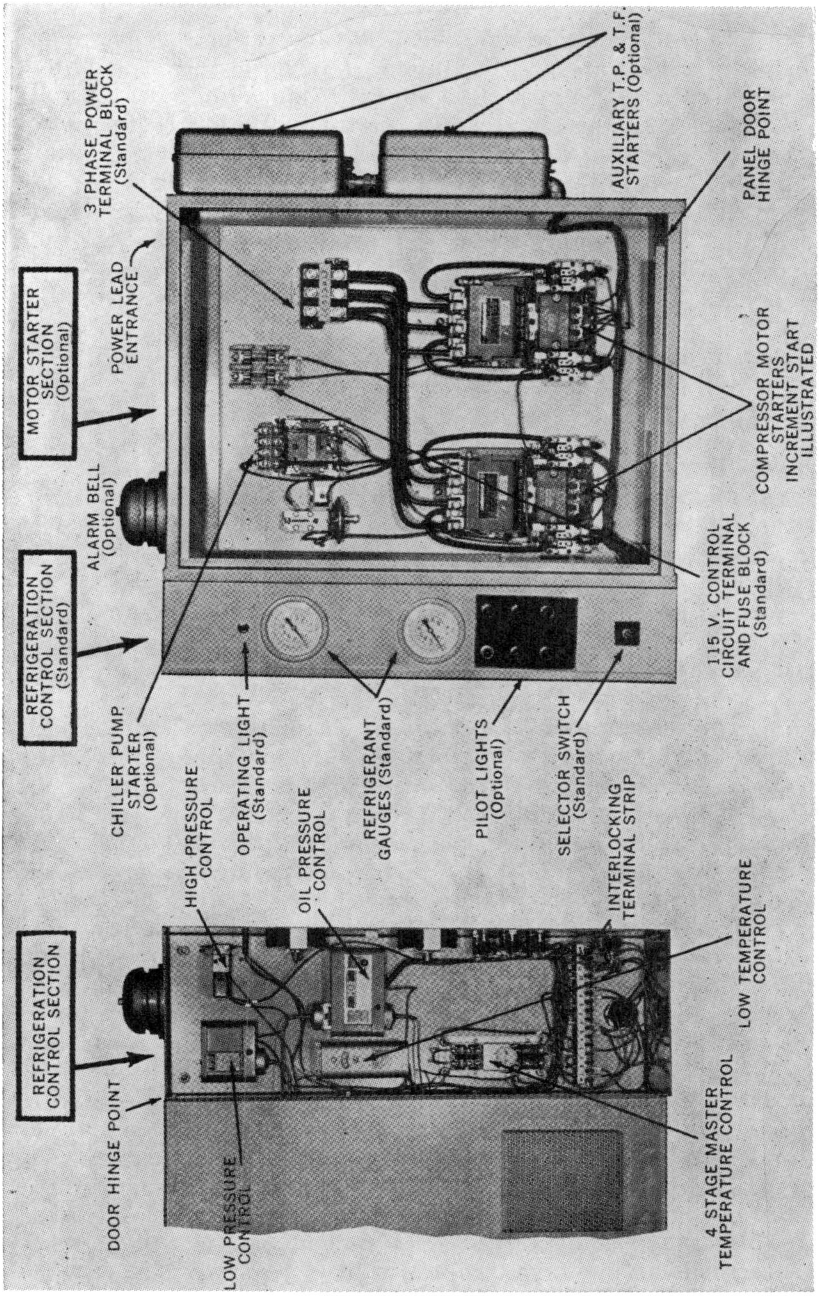
REFRIGERATION CONTROL SECTION
DOOR HINGE POINT
LOW PRESSURE CONTROL
HIGH PRESSURE CONTROL
OIL PRESSURE CONTROL
INTERLOCKING TERMINAL STRIP
LOW TEMPERATURE CONTROL
4 STAGE MASTER TEMPERATURE CONTROL
REFRIGERATION CONTROL SECTION (Standard)
CHILLER PUMP STARTER (Optional)
OPERATING LIGHT (Standard)
REFRIGERANT GAUGES (Standard)
PILOT LIGHTS (Optional)
SELECTOR SWITCH (Standard)
ALARM BELL (Optional)
MOTOR STARTER SECTION (Optional)
POWER LEAD ENTRANCE
3 PHASE POWER TERMINAL BLOCK (Standard)
AUXILIARY T.P. & T.F. STARTERS (Optional)
PANEL DOOR HINGE POINT
COMPRESSOR MOTOR STARTERS INCREMENT START ILLUSTRATED
115 V. CONTROL CIRCUIT TERMINAL AND FUSE BLOCK (Standard)

Fig. 41

Fig. 42

called for by the temperature controller. This control prevents operation due to loss of refrigerant and expansion valve failures, etc. and shuts down the unit after a "pump-down" after the liquid line solenoid valve closes in response to the temperature control indicating cooling is no longer called for.

(b) High Pressure Cut-Out. This is strictly a safety device to shut down the unit due to high refrigerant discharge pressure. This can be caused by lack of condensing water, high temperature condensing water, a refrigerant overcharge, closed valves or similar malfunctions.

This is a lockout type control and must be manually reset to restart the unit.

(c) Anti Freeze Control. This control shuts the unit down if the temperature of the chilled water approaches freezing temperature. It prevents the freezing of water in the chiller when properly adjusted. It is a safety control and should not operate during normal operation. It may or may not be of the lockout type requiring manual reset.

(d) Oil Pressure Safety. This is a safety device of the lock out type which shuts down the system due to the lack of oil pressure required for compressor lubrication.

It includes a switch which opens when the oil pressure is sufficiently above suction pressure to assure proper compressor lubrication. This switch is in series with a heating element which causes a thermally operated switch in the control circuit to open after a desirable time delay.

If the control circuit calls for cooling, a small heater simultaneously begins to heat. If the compressor starts, and the oil pressure builds up within a prescribed time, the pressure difference switch turns off the heater and the system continues to operate normally. However, if the system starts and the oil pressure does not build up normally, the heater continues to heat until it opens a thermally operated switch in the control circuit and shuts down the system.

(e) Temperature Controller. A multiple step temperature controller is used to control compressor capacity, by causing progressive cylinder unloading, as cooling demand decreases, as indicated to a thermal bulb in the return chilled water line. As the final stage is reached the liquid line solenoid valve is closed, causing the evaporator to pump down until the low pressure control opens and stops the motor.

A separate electrical panel Fig. 42 contains the motor starters used to energize the first and second increment windings of the compressor motor and such other relays as may be required. For a normal fully interlocked system the chilled water pump is interlocked to insure that the chiller does not refrigerate unless water is circulating through the chiller. This can be accomplished by a contact on the chilled water pump starter.. It can also be accomplished by using a water flow switch which closes only if it is activated by the chilled water flow.

Cooling tower pump and fan motors may also be interlocked to the extent that if any motor starter is deenergized due to an overload, the open overload contacts will shut down the entire system.

The control panel as shown by Fig. 41 is attached to the motor starter panel. The motor starter section, Fig. 42, houses the first and second increment compressor motor starters, the chiller pump starter, a time delay, a fuse block for the 110 volt control wiring and the 3 phase power terminal block. Attached to the right side of the starter panel, two auxiliary starters are secured which are interlocked into the control wiring and which are used to operate the cooling tower pump and fan motors.

The electric package liquid chiller for use with a cooling tower is a complete "package" requiring only electrical and water piping connections at installation. The power supply leads must be brought to the starter panel, and leads from the chiller pump, tower fan, and tower pump, starters are wired to their respective motors. Condensing water piping in and out is connected to the condenser and chilled water supply, and return piping to the chiller is also field connected.

Liquid Chiller Controls

Electric package liquid chillers in sizes up to 200 hp are entirely automatic machines and do not require the services of an operating engineer although periodic inspection and maintenance is recommended. In order to provide entirely automatic operation a number of safety controls are required.

To provide completely automatic operation of an entire system it is necessary to interlock the action of cooling tower fan and pump motors and the chilled water pump with the controls of the package liquid chiller. Also there may be some other remote controls such as a summer-winter switch, a remote off-on switch, or even some other safety control which is connected on the job. Jumpers can be removed from the factory wired control system for easy connection of such field wired controls.

Electric package liquid chillers are available, using a single open type compressor and compressor motor in sizes from 25 to 100 hp and using four steps of capacity control with the control wiring arrangement as shown by Fig. 43.

Automatic controls are mostly temperature operated or pressure operated switches, time delays, and relays or motor starters. The schematic symbols shown in Fig. 46 indicate the type of control device in the diagram.

1. Preliminary requirements

(a) Control line must be energized from 115V separate circuit or from transformer with primary connected across one phase of 3 phase

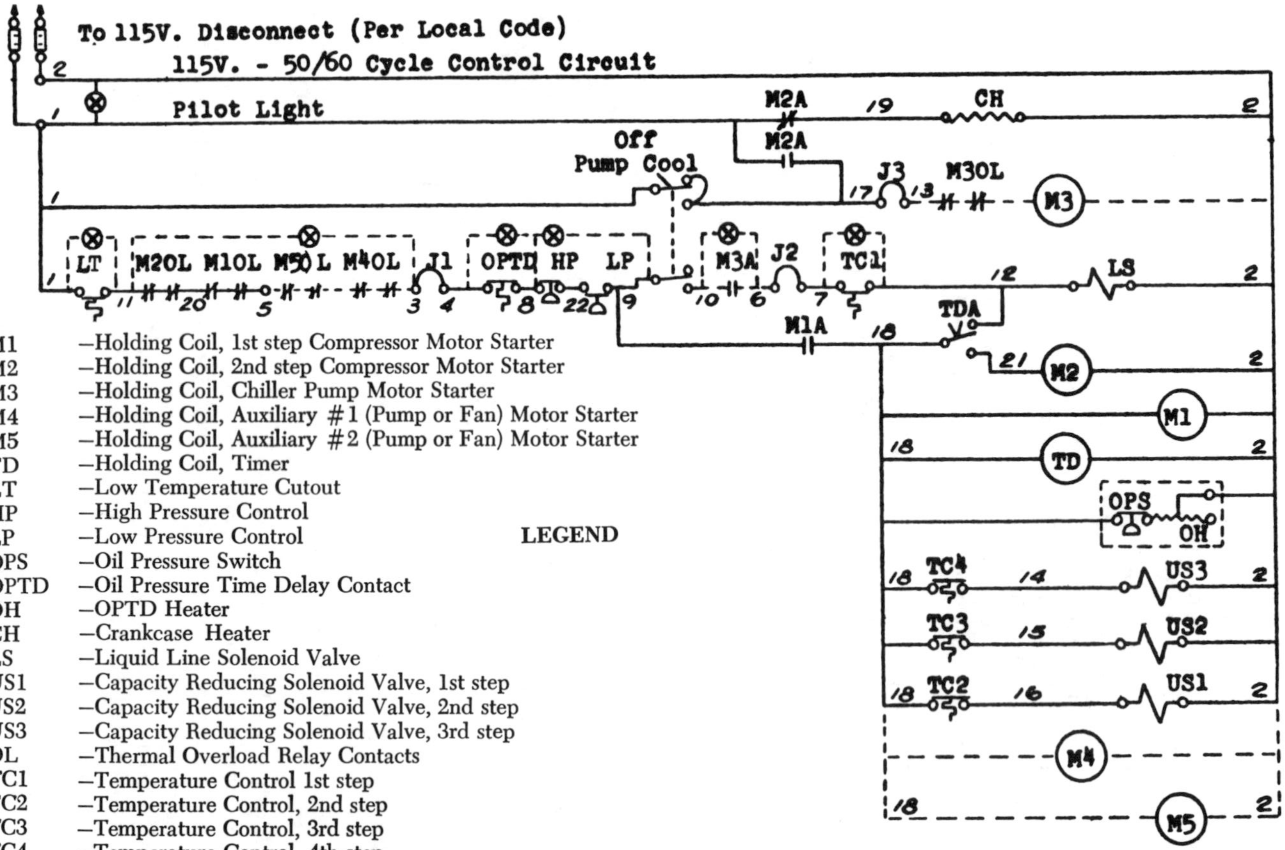

Fig. 43 — Single compressor package chiller with four step capacity control.

LEGEND

M1	—Holding Coil, 1st step Compressor Motor Starter
M2	—Holding Coil, 2nd step Compressor Motor Starter
M3	—Holding Coil, Chiller Pump Motor Starter
M4	—Holding Coil, Auxiliary #1 (Pump or Fan) Motor Starter
M5	—Holding Coil, Auxiliary #2 (Pump or Fan) Motor Starter
TD	—Holding Coil, Timer
LT	—Low Temperature Cutout
HP	—High Pressure Control
LP	—Low Pressure Control
OPS	—Oil Pressure Switch
OPTD	—Oil Pressure Time Delay Contact
OH	—OPTD Heater
CH	—Crankcase Heater
LS	—Liquid Line Solenoid Valve
US1	—Capacity Reducing Solenoid Valve, 1st step
US2	—Capacity Reducing Solenoid Valve, 2nd step
US3	—Capacity Reducing Solenoid Valve, 3rd step
OL	—Thermal Overload Relay Contacts
TC1	—Temperature Control 1st step
TC2	—Temperature Control, 2nd step
TC3	—Temperature Control, 3rd step
TC4	—Temperature Control, 4th step
J1	—Tie in for field wired safety control
J2 and J3	—Tie in for field wired operating (remote) control

power circuit with cartridge fuses as shown. Circuit is energized when signal light is "on".

(b) Three phase power is connected through closed, fusel, disconnect switch to terminals L1, L2, and L3 in starter panel box.

(c) The anti-freeze control LT is closed, the overload contacts M20-L, M10L, M50L, and M40L are closed in the second increment compressor motor starter, the first increment motor starter, the cooling tower pump starter, and the cooling tower fan starter, respectively.

(d) The compressor oil pressure time delay OPTD switch is closed, the refrigerant high pressure cut out switch HP is closed and the low pressure control LP is closed.

2. Set Selector Switch to "Cool"

(a) In the "cool" position a two pole three position switch upper contact connects terminal 1 to terminal 17 and through overload contacts M30L and the holding coil M3 in the chilled water pump starter to terminal 2. This causes the chilled water pump to start and also closes contact M3A.

(b) In the "cool" position the lower pole contact completes the circuit from terminal 9 to 10 and through contact M3A, J2, temperature controller contact TC1 and energizes the liquid line solenoid valve LS allowing liquid refrigerant to feed to the expansion valve.

(c) With terminal 12 energized a circuit is completed through time delay contact TDA to terminal 18. This energizes the first increment motor starter M1 starting the compressor motor. It also energizes the time delay TD, the oil pressure safety heater OH, holding coils M4 and M5. At high water temperatures (60+) temperature controller switches TC4, TC3, and TC2 should be open. Also with M1 closed contact M1A closes to energize terminal 18 independent of the selector switch, contact M3A, jumper J2 and temperature controller contact TC1.

(d) After TD has been energized for 1 to 3 seconds contact TDA will shift to energize M2 the second increment winding of the compressor motor so the motor is fully on the line and the compressor crankcase heater is de-energized by the opening of contact M2A (NC) while N2A (NO) is closed.

(e) If the compressor oil pump raises the oil pressure within seconds after the compressor starts it will open the oil pressure switch OPS and de-energize the heater OH. The oil pressure will also operate the cylinder loading mechanism in all cylinders to fully load the compressor. With M4 and M5 energized, the cooling tower and chiller are in full operation.

(f) If the compressor oil pressure failed to rise when the compressor started the cylinders would not load as loading is responsive to oil pressure. Also the switch OPS would not open and the heater OH would continue to be energized to heat the thermostatic switch in the safety control OPTD until it opened and shut down the compressor motor. This would keep the compressor shut down but would leave the chilled water pump operating. A manual reset button in the OPTD control would have to be reset in order to put the compressor back in operation and thereby warns against oil pressure failure.

Manual Shutdown Sequence

(a) To shut the system down manually, the selector switch is moved from the "cool" to the "off" position in which case neither pole of the switch completes a circuit and terminal 12 is de-energized causing the liquid line solenoid valve LS to close.

(b) After the suction pressure is reduced by "pumping down" the evaporator, the low pressure control opens and de-energizes line 18 shutting down the cooling tower motors and compressor motor. Also de-energizing holding coil M2 will cause contact M2A to open the normally open contact and shut down the chilled water pump while the normally closed contact M2A will re-energize the crankcase heater CH to prevent refrigerant absorbtion in the crankcase oil.

(c) After the compressor has been shut down, residual refrigerant in the evaporator will cause a partial suction pressure build up sufficient to reclose the low pressure control. However, the selector switch will have to be reset to "cool" to start the compressor again as contact MIA opened when M1 was de-energized and contact TDA returned to the position shown when TD was de-energized.

Automatic Start and Load Sequence

(a) If the compressor and cooling tower motors have shut down due to satisfying the load the chilled water pump continues to run as M3 is energized by the selector switch which remains in the "cool" position. Temperature control contacts TC2, TC3, and TC4 are closed during automatic shut down and TC1 is open.

(b) With an increase in return chilled water temperature TC1 will close. This opens the solenoid valve LS and energizes terminal 18 through TDA which energizes M1 which in turn energizes MIA and the first increment motor winding, M4, M5, TD, unloader solenoid valves US3, US2, and US1, also heater OH. Consequently, the compressor motor and the cooling tower motors start. After 1 to 3 seconds TD switches contacts at TDA and the second increment of the compressor motor is on the line.

(c) After the compressor oil pressure builds up, switch OPS opens and at least 25% of the compressor cylinders (not connected through unloader solenoid valves) load and the machine operates at least 25% capacity.

(d) If 25% capacity is sufficient to reduce the return chilled water temperature about 2°, TCI will open and cause a normal compressor shut down by closing the liquid solenoid valve and pumping down the evaporator until the low pressure control stops the compressor and tower motors.

However, if the load increased until the 25% capacity was so low as to allow TC2 to open, solenoid valve US1 would close and cause another 25% of the cylinders to load so that the machine would operate at 50% capacity.

(e) With further return water temperature increases TC3, and TC4, would open progressively closing US2, and US3 with each step loading another 25% of the compressor until full capacity is reached.

Automatic Unload and Shut down Sequence

As return chilled water temperature progressively decreases, indicating a reduced load the following events take place.

(a) TC4 closes. US3 opens. 25% capacity reduction.

(b) TC3 closes. US2 opens. Capacity reduced to 50%.

(c) TC2 closes. US1 opens. Capacity reduced to 25%.

(d) TC1 opens. LS closes. Evaporator pumps down, LP. opens and compressor and tower motors shut down. Chilled water pump continues to run. Crankcase heater energized when compressor motor is shut down.

Control Panel

The control panel diagram included with the machines, shows the terminal numbers and controls included in this panel which includes the low pressure control LP, the high pressure cut-out HP, the anti-freeze control AF with a capillary tube and bulb located in the chilled water outlet, the master temperature control containing 4 steps and with capillary tube and control bulb in the return chilled water pipe, the two pole 3-position selector switch and the required terminal strips.

The increment start panel with auxiliary starters contains the power line entrance terminals (L1, L2, L3) the first increment motor starter M1, the second increment motor starter M2, the fused 115 volt control power entrance block, the chilled water pump starter M3 and the time delay inside the panel. The cooling tower fan and pump motor starters M3 and M4 are not shown in Figs. 43 and 45 mounted on the outside of the panel. All interconnecting wires have numbers attached to them to indicate the terminal numbers to which they should be attached.

Dual Units

Fig. 44 shows a dual compressor unit which ITT Bell & Gossett manufactures in sizes from 60 to 200 hp. The inlet and outlet chilled water connections are shown extending through the insulation around the chiller. Two compressors, two compressor motors, and two condensers are used. The chiller has only one water circuit but has two separate refrigerant circuits so that each motor, compressor, condenser, and refrigerant controls with its section of the chiller constitutes a complete refrigeration system in itself. If desired, one condensing unit can be serviced while the other is in operation.

The controls for the dual unit chiller are necessarily more involved than for a single compressor system especially where the order in which the two machines are started is changed in each succeeding cycle. Also provisions must be made so that the two compressor motors cannot be started at the same time which would take a tremendous current demand during starting.

Fig. 45 shows the control wiring for a dual compressor system. In order to simplify an apparently complicated control arrangement each of several control circuits can be considered one at a time. The control diagram shows 115 volt control power supplied by a transformer with its primary leads connected to one phase of the 3 phase power lines. Line 2 is a common line to which all circuits must be completed from line 1 to be operative.

Each of 11 circuits are marked to aid in a discussion of the circuits. The function and operation of each of these is as follows:

Circuit 1. This circuit from line 1 has contact M2A (N.C.) in series with CH. M2 is the holding coil for the second increment motor starter for No. 1 compressor and M2A is a contact on the starter. CH is the crankcase heater in No. 1 compressor which is energized as long as the control circuit is energized and the compressor motor is not operating but is de-energized by contact M2A when the compressor motor started is closed.

Circuit 2. This is a duplicate of circuit 1 except that contact M7A energizes the No. 2 compressor crankcase heater when the compressor is not in operation and de-energizes the heater when the compressor operates.

Fig. 44 — Dual compressor unit.

Circuits 3 and 4. Contact R1A closes when R1 is energized indicating that No. 1 compressor is operating. Contact R2A is closed when R2 and No. 2 compressor motor is energized. In parallel they serve to operate tower fan and pump motors whenever either unit 1 or unit 2 is in operation through motor starters M4 and M5.

Circuits 5 and 6. These circuits serve to provide a selective setting through a two pole three position switch to control the lead and lag starting arrangement for the two condensing units.

(a) With the selector switch at No. 1 the upper pole connects to an open contact and the lower pole connects terminal 1 to 30 which causes relays R3 and R4 to be energized. With this condition all R3A, R3B, R3C, R3D, R4A, R4B, R4C, and R4D normally open contacts will close and all of these normally closed contacts will open. Under this condition No. 1 compressor will start first, as the temperature control contact TC1 closes and loads half of No. 1 compressor. The second half of No. 1 compressor will load after control TC2 opens.

Number 2 compressor will start after control TC3 closes and time delay TD3 has closed contact TD3A connecting terminals 38 to 44. The compressor will load 50% when its oil pressure builds up.

When TC4 opens No. 2 compressor will load to 100% capacity.

(b) With the lead-lag switch set at No. 2 both upper and lower poles will be open and relays R3 and R4 will not be energized so when TC1 closes No. 2 compressor will start and load 50% when its oil pressure builds up and will load to 100% capacity when TC2 opens. No. 1 compressor will start and load to 50% capacity when TC3 closes and its oil pressure builds up. It will load to full capacity when TC4 opens. This action is the same as before but with the order of the compressor starting reversed.

(c) With the lead-lag switch in the "Alt" position as shown, AR is de-energized every time that contacts R1A or R2B open or every time that either compressor operates. AR is the operating solenoid for the switch from terminals 29 to 30 which opens and closes with each succeeding operation of the system.

When the switch is closed R3 and R4 are energized and No. 1 compressor starts first. If the switch is open R3 and R4 are de-energized and No. 2 compressor starts first.

If No. 1 compressor handles the load alone and shuts down without No. 2 unit operation, No. 2 unit will start first the next time cooling is called for and vice versa when the lead-lag switch is in the position shown.

Circuit 7. This circuit is fed by normally open contacts M2A and M7A which are auxiliary contacts on No. 1 motor second increment and No. 2 motor second increment starters respectively so that terminal 17 will be energized when either motor is operating. This insures that M3 will keep the chilled water pump operating as long as either

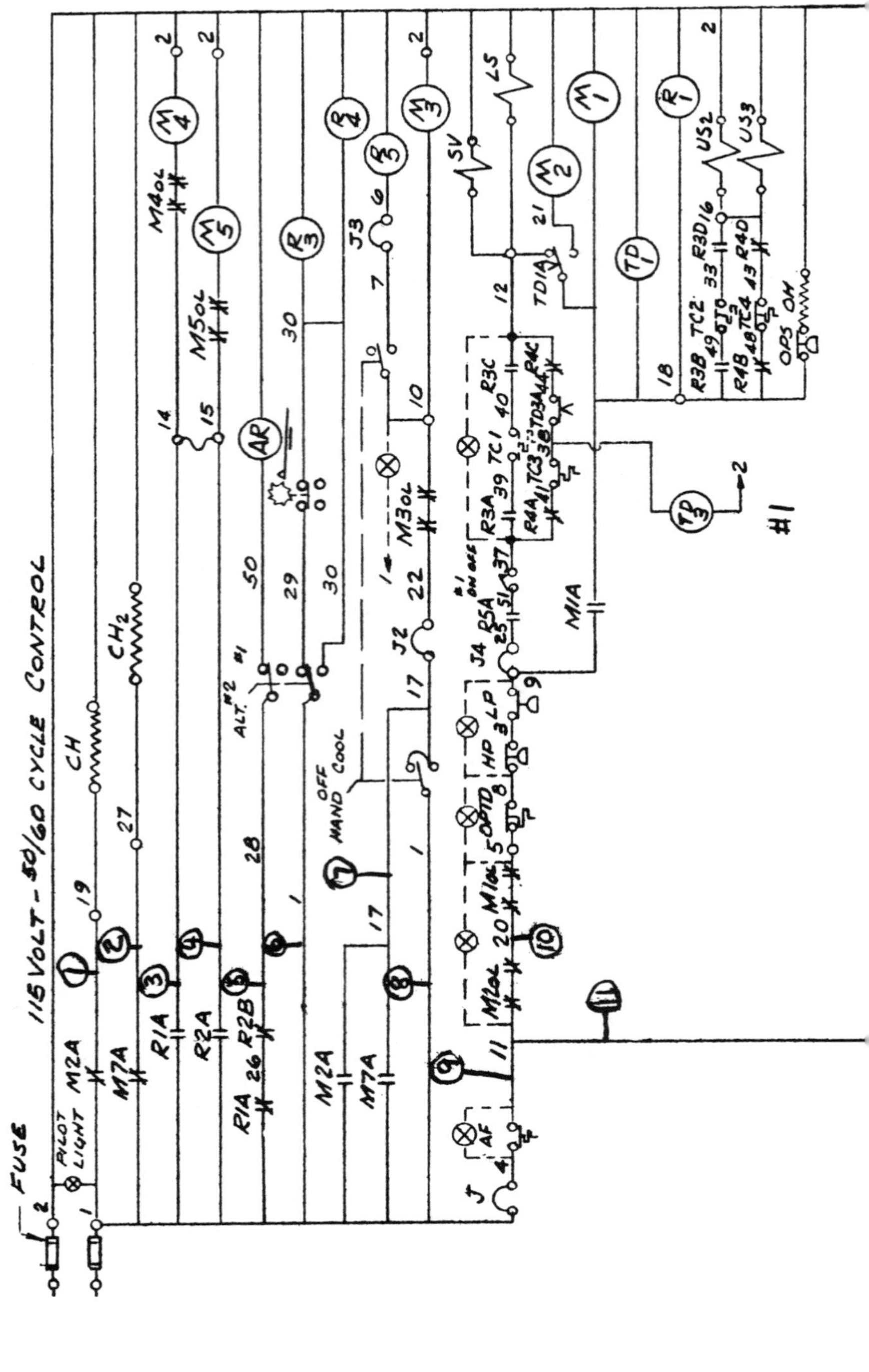

115VOLT - 50/60 CYCLE CONTROL
FUSE
PILOT LIGHT
M2A
M7A
R1A
R2A
R2B
M2A
M7A
CH
CH2
M4OL
M5OL
M3OL
M1OL
M2OL
HAND OFF COOL
ALT.
AR
SV
LS
TD1A
TC1
TC2
TC3
TC4
R3A
R4A
R3C
R4C
R3B
R4B
R3D
R4D
R5A
M1A
US2
US3
OPTD
HP
LP
AF
OPS OH
J
J2
J3
J4
#1

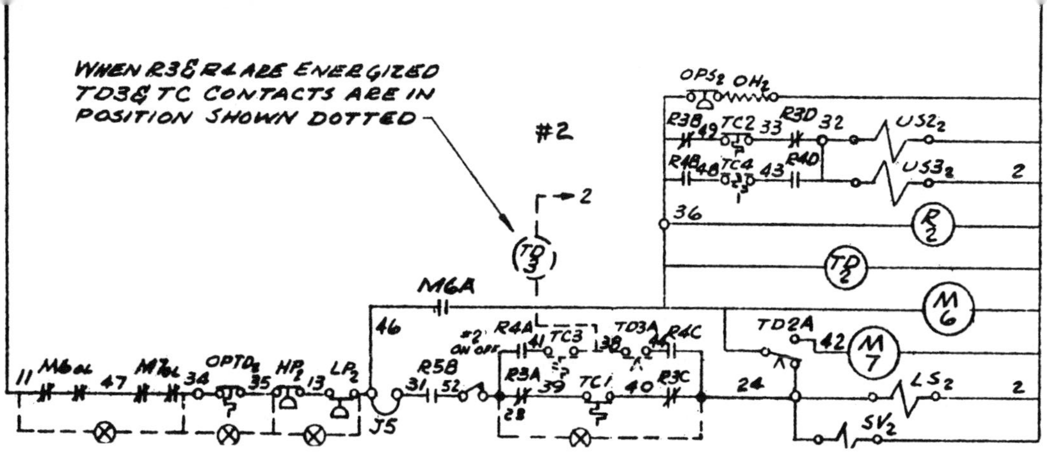

Fig. 45 — Schematic for dual compressor and legend.

M1 and M6	—Holding Coil, 1st step Comp. Mtr. Starter
M2 and M7	—Holding Coil, 2nd step Comp. Mtr. Starter
M3	—Holding Coil, Chiller Pump Mtr. Starter
M4	—Holding Coil, #1 Auxiliary Mtr. Starter
M5	—Holding Coil, #2 Auxiliary Mtr. Starter
R1-R5	—Holding Coil, Control Relay
TD1-TD3	—Holding Coil, Time Delay Relay
AR	—Holding Coil, Alternator Relay
AF	—Anti-Freeze Control
HP	—High Pressure Switch
LP	—Low Pressure Switch
OPS	—Oil Pressure Switch
OPTD	—Oil Pressure Time Delay
OH	—OPTD Heater
CH	—Crankcase Heater
OL	—Thermal Overload Relay Contacts
LS	—Liquid Line Solenoid Valve
US	—Unloader Solenoid Valve
SV	—Bleed-Off Solenoid Valve
TC	—Temperature Control (4 Stage)
J	—Tie in for field wired Safety Control(s) (flow switch, etc.)
J2	—Tie in for field wired Remote Control (units and chiller pump)
J3	—Tie in for field wired Remote Control (units only
J4, J5	—Tie in for field wired Remote Control (units individually).

compressor operates when the selector switch has been moved from "cool" to "off" to effect manual shut down of the system. The chilled water pump will shut down when both compressors stop with the selector switch at "off".

Circuit 8. This circuit energizes M3 when the selector switch is in the "cool" position causing the chilled water pump to operate continuously independent of compressor operation. The selector switch in the "cool" position also energizes relay R5 which closes contacts R5A and R5B which is required for units No. 1 and No. 2 to start in response to the closing of temperature control TC1. When the selector switch is moved from "cool" to "off" relay R5 is de-energized and will not allow either compressor to restart after it has been stopped by its low pressure control on a pump down operation even though the pressure control recloses and the temperature control calls for cooling.

Circuit 9. This includes jumper J1 and the anti-freeze control AF. In some cases a flow switch which is closed by the flow of water in the chilled water line, into which it is connected, replaces jumper J1. This will prevent either compressor from operating if the chilled water temperature goes below a safe lower limit, or if the chilled water flow rate is below the required flow.

Circuit 10. This is the control circuit for compressor No. 1 and involves its own safety controls, liquid solenoid valve, low pressure control, etc. It is essentially the same as shown by Fig. 43 except that it involves the double throw contacts which in the position shown will start the No. 1 compressor when temperature control TC3 closes, and loads it to 50% capacity and will load it fully when TC4 opens. If TC1 and TC3 were both closed and TC2 and TC4 were both open, the temperature control would be calling for full capacity of both compressors. If the selector switch was turned to "cool" the controls would call for full capacity of both compressors at once. Time delay TD3 will hold contact TD3A open and prevent No. 1 compressor from starting until No. 2 unit has started and loaded.

Circuit 11. This circuit contains the individual controls and safeties for compressor No. 2. With all R3 and R4 relay contacts in the position shown, TC1 would start this compressor and TC2 would fully load it. This compressor motor would start immediately when the selector switch is turned to "cool" if all temperature controls called for maximum cooling. If the position of all R3 and R4 contacts were reversed No. 2 compressor would not start until No. 1 compressor had started and loaded.

When two or more condensing units are connected to separate circuits of a chiller there is the possibility that one unit may handle the load adequately for hours or days without causing the second compressor to start. In this case any leakage through the compressor discharge valves or through the liquid line solenoid valve of the non-operating unit would tend to gather as liquid refrigerant in the evaporator since it would be the coldest point in the system especially since it is maintained at low temperature by the operating unit. This can cause liquid slugging when the capacity demand finally causes the second compressor to start. This liquid slugging is alleviated by installing a liquid trap in the suction line.

The liquid from the trap is drained into the suction line through a solenoid valve and small line at such a rate that it can be tolerated by the compressor. Drainage into the compressor during the compressor shut down is prevented by the closing of the solenoid valves SV in parellel electrically with the liquid line solenoid valves so that drainage occurs only when the compressor is operating.

Alternate Arrangements

Other manufacturers of package liquid chillers manufacturing similar equipment may use other control combinations and in each case the control diagram referring to the particular unit should be used when control servicing is required. In some cases capacity variation in 25% steps is called for in engineering specifications. This can be accomplished also by the use of four complete condensing units each serving one of four evaporator sections. In such cases capacity control can be effected by stopping a compressor rather than unloading part of its cylinders.

Essentially the same major control functions are required in any case and these may be summarized as follows:

(a) The chiller must be protected against freeze up. This can be accomplished by a low temperature refrigeration thermostat responding to chilled water temperature which will immediately shut the compressor down when it opens as in the diagrams shown. It can also be installed in series with the liquid line solenoid valve so that it would cause the evaporator to pump down and shut down as in a normal shut down cycle. This type control recloses when the temperature rises.

(b) With motors having two increment windings and using two starters to energize the windings, it is necessary to shut the system down if either winding is overloaded. The overload cut-outs require manual reset to indicate that the cause of the overload should be located and corrected.

(c) Failure to build up oil pressure for compressor lubrication could be serious in two ways. First the lack of bearing lubrication could cause compressor damage and second, failure to load at least one compressor cylinder would allow the compressor to operate 100% unloaded and the frictional heat in the gas will build up the temperature rapidly if there is no gas circulation through the compressor to carry the heat away. The time delay between compressor start up and shut down due to oil pressure failure should not be too short or many service shut downs may occur which would be prevented by a longer time allowance. In some instances oil may migrate from the crankcase to other parts of the system at shut down and require a short interval to return to the crankcase at start up which can delay oil pressure build up.

The pressure switch which de-energizes the heater which causes the shut down if the oil pressure fails to build up is a differential pressure switch and responds to the oil pressure difference above suction pressure. It is a lockout type sswitch that requires manual resetting at which time the reason for the lack of oil pressure should be corrected.

(d) A refrigerant high pressure cut out can open due to excessive pressures caused by lack of condensing water, refrigerant overcharges, fouling of the condenser heat transfer surface, etc. This also is a

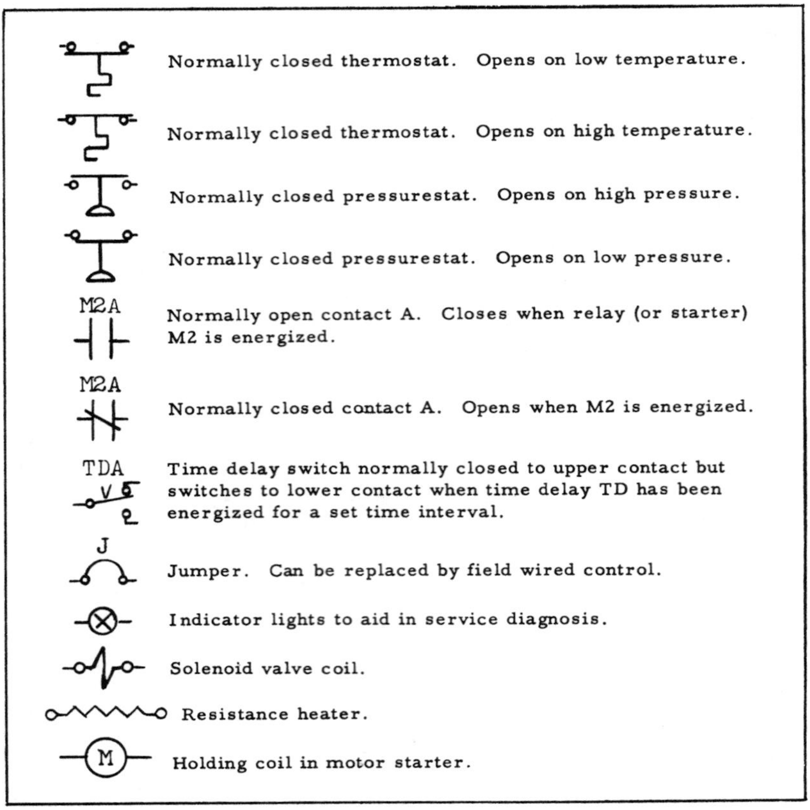

Fig. 46 — Symbols.

lockout control requiring manual reset when the cause of the high pressure should be located and corrected.

(e) Liquid slugging is almost certain to occur if the compressor operates without a load on the liquid chiller so it is necessary that adequate water flow through the chiller be maintained. This is the reason for the control system providing that the compressor cannot operate unless the chiller pump starter is closed. A flow switch provides still better protection as it is possible for the chilled water pump starter to be closed and still not provide adequate chilled water flow. The control systems also provide a means of keeping the pump operating during pump down when the selector is in a manual shut off position.

Essentially the above 5 safety functions must be provided. Capacity control arrangements may vary according to compressor design and the experience of the design engineer or even the particular installation the chiller is to supply.

Indicator Lights

A power-on light shows when 115 volt power is supplied to the control system. Optional indicator lights may be provided to indicate the location of a service fault or the cause of compressor shut down. Labels at the signal locations readily locate the reason for compressor shut down as:

1. Temperature control open. No trouble if temperature O.K.
2. Motor overload contacts open
3. Chiller pump motor starter open
4. High or low pressure control open
5. Compressor oil pressure failure
6. Anti-freeze control open

Liquid Chiller Installation

The installation of a 25 to 200 hp package liquid chiller requires preliminary planning. If the chiller is to be installed in a new machinery room, a solid concrete foundation separate from the floor is recommended. The weight of the reinforced concrete should be about 2½ times the weight of the chiller assembly. A foundation of this nature is shown in Fig. 47. When this type foundation is provided the machine can be mounted on the foundation and leveled by means of shims. A grout of sand and cement or a prepared grout material should be flowed into the space around the shims and mounting bolt holes. With a solid mounting of this nature the condenser and chilled water piping can be directly connected to the appropriate connections.

Where the machine is to be mounted on an existing concrete floor over the ground an additional base can be prepared as in Fig. 48. For mounting on concrete floors above grade level, a rubber in shear type vibration eliminator as recommended by the manufacturer may be used. If necessary the weight of the machine can be spread over more area by timbers under the mounting points to prevent exceeding the allowable floor loading. For mounting on wood floors, when necessary, spring type resilient mounts may be used. The services of an engineer specializing in vibration problems may be a good investment in special cases.

Wherever the chiller is mounted on a resilient mount it is free to move slightly with respect to the floor on which it is mounted. Con-

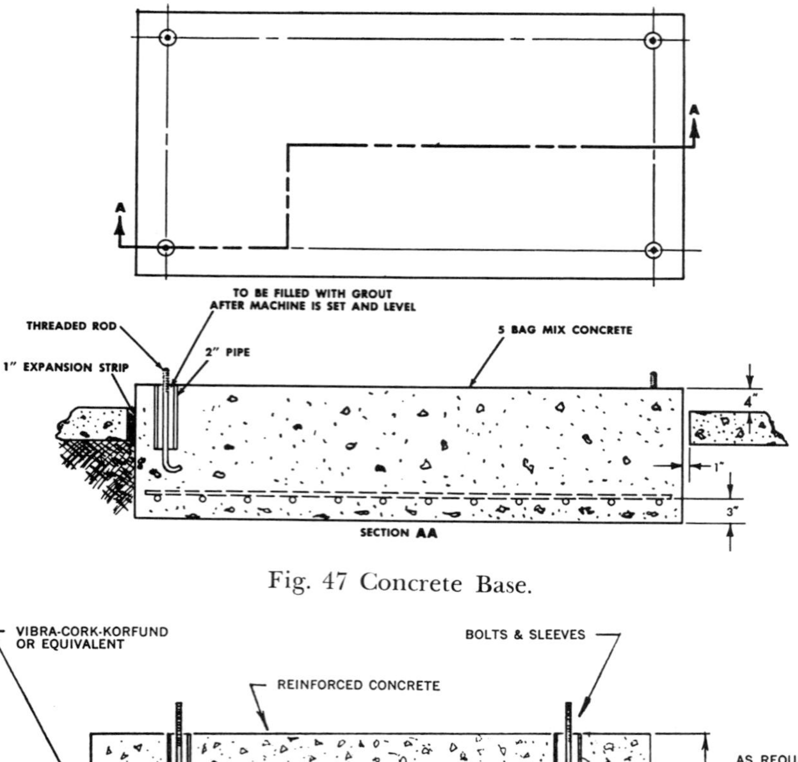

Fig. 47 Concrete Base.

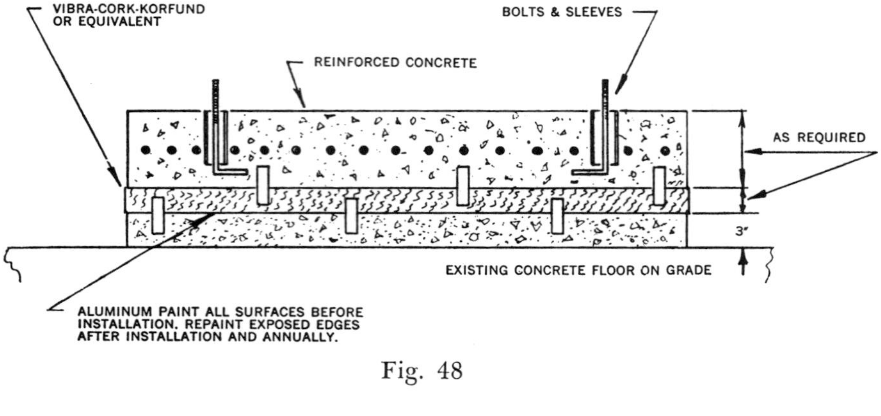

Fig. 48

sequently flexible piping connections should be provided. The building system piping should be anchored to a solid support near the chiller and flexible piping should be provided between the chiller and the anchor point. If a single flexible connection is used, it should have ample length to absorb lateral as well as transverse vibration. In some cases two short flexible connections at right angles may be used to absorb vibration in two directions.

Chilled Water Piping

The condenser cooling and chilled water connections may be bushed down at the condenser and chiller if desired to a pipe size which will provide a velocity which would cause a piping pressure drop not to exceed 500 to 600 mil inches per foot. See Fig. 15.

The rate of flow for the chilled water piping in gpm will be:

$$\text{Gpm} = \frac{24\ \text{T}}{^\circ\text{TD}}$$ where T is tons of capacity and °TD is the temperature drop of the water through the chiller, usually 8° to 12° at full load.

Condensing Water Piping

For "city water" or cooling tower water the heat rejection by the condenser will be approximately 15,000 Btu per hr per ton and the condenser flow rate in gpm is:

$$\text{Gpm} = \frac{30\ \text{T}}{^\circ\text{TD}}$$ where T is tons of capacity and °TD is the temperature rise of the condensing water.

Refrigerant Piping

Where water is so scarce it may not be available in sufficient quantities to be used as make-up water for cooling tower use it becomes advisable to use an air cooled condenser. In such cases the package liquid chillers are ordered without condensers and air cooled condensers are field installed.

The selection of the hot gas piping should be made with regard to the problem of carrying the oil up vertical lines. It is desireable to install lines large enough to keep pressure loss low but to maintain velocity high enough in the partial load conditions so that the oil will carry along with the refrigerant in vertical lines.

Fig. 49 shows a piping selection chart for R-22 discharge lines. Dashed lines indicate selection procedure for an R-22 discharge line for a 100 ton refrigeration system with unloading down to 25 tons, with a condensing temperature of 130°F.

Procedure: At lowest load determine line size for vertical risers which will maintain oil circulation.

(a) On right side of chart estimate 25 tons capacity. (A)

(b) Follow left parellel to 20 and 30 ton capacity lines to 130° vertical condensing intersection (B)

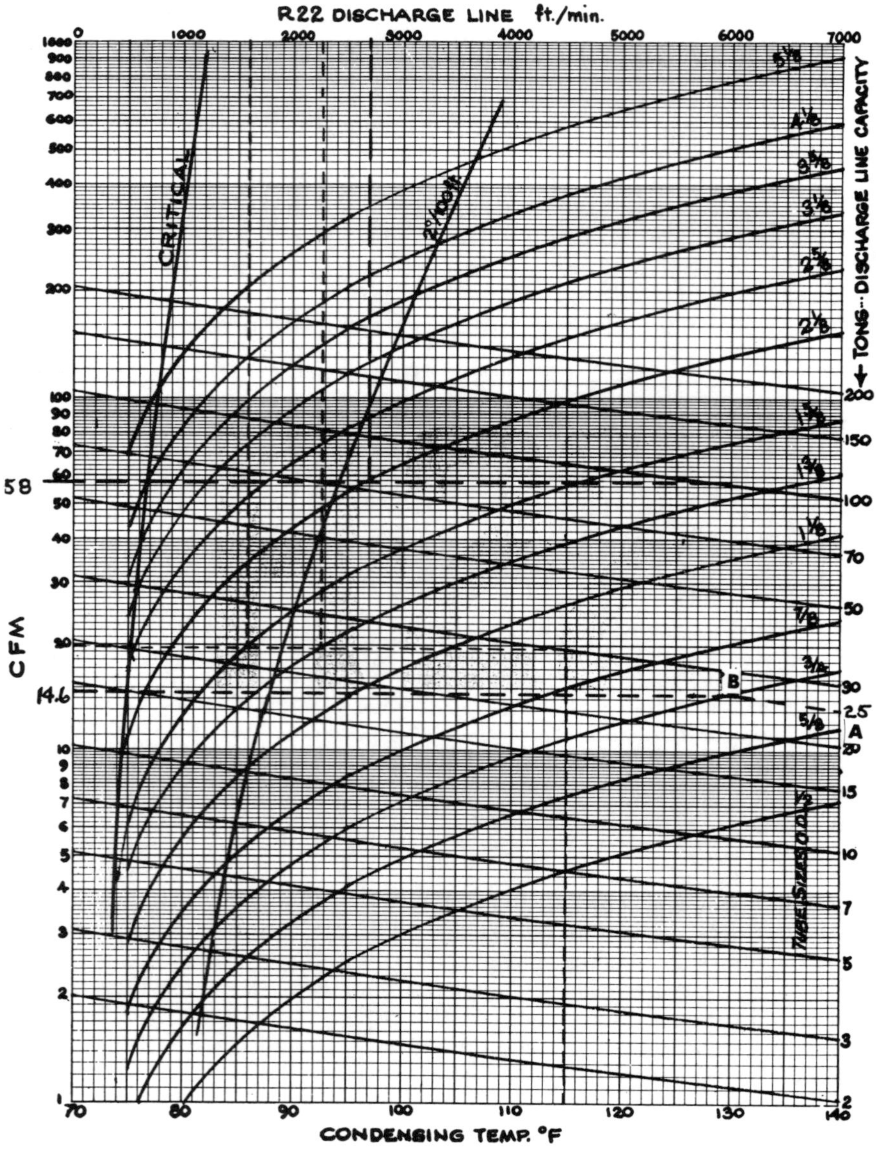

R22 DISCHARGE LINE ft./min.
0
1000
2000
3000
4000
5000
6000
7000
CRITICAL
2°/100 ft
5 1/8
4 1/8
3 5/8
3 1/8
2 5/8
2 1/8
1 5/8
1 3/8
1 1/8
7/8
3/4
5/8
1/2
TUBE SIZES (O.D.)
TONS – DISCHARGE LINE CAPACITY
CFM
58
14.6
B
A
2.5
CONDENSING TEMP. °F

Fig. 49

(c) Follow horizontally to left to find cfm hot gas flow rate at about 14.6 cfm.

(d) From intersection of 14.6 cfm and critical line read critical velocity at top of chart as 500 fpm.

(e) The 2⅝ tube line curve intersects the 14.6 cfm line at a velocity less than 500 fpm while the 2⅛ tube curve intersects the 14.6 cfm line at about 660 fpm which will assure oil circulation up vertical risers at minimum load.

(f) At 100 tons and 130° condensing the flow rate is indicated as about 58 cfm and the intersection with the critical line shows about 680 as the critical velocity whereas in a 2⅛ OD tube the velocity will be about 2700 fpm. The 2⅛ OD tube will give a pressure drop of about 8 psi per 100 ft of equivalent length of 2⅛ tube and about 4 psi per per 100 ft in a 2⅝ line. If the compressor minimum load was 1/3 of full load the 2⅝ OD discharge line would be satisfactory. If the discharge line is only 50 ft long the pressure drop of 4 psi would represent a compressor apparent condensing temperature of between 1 and 2° above the actual condensing temperature.

Fig. 50 shows similar discharge selection curves for R-12. It will be noted that a 2⅝ line should be used for an R-12 100 ton system unloading to 25 tons at 130°.

Where horizontal sections of discharge lines are required one size larger line can be used in the horizontal runs but the vertical lines should be 2⅛ OD for R-22. As the condensing temperature is reduced the line size may be increased.

Liquid Lines

There is no trouble with oil circulation when mixed with liquid refrigerant. The problem with liquid circulation involves a means of insuring that the liquid refrigerant is subcooled at the expansion valve. One of the best means of obtaining this is to locate the liquid receiver a short distance below the condenser and preferably several feet at least above the expansion valves. This provides a static head which more than offsets any pressure loss due to friction in the line.

In general if the liquid level in the receiver is 10 feet or less above the expansion valve a liquid line selected to give a pressure drop of 5 psi per 100 feet will be satisfactory unless the horizontal lengths are over 5 times the vertical drop.

For receivers over 15 feet above the expansion valves, and the vertical drop exceeds the horizontal run, lines selected on the basis of 10 psi loss per 100 feet will be satisfactory.

Lines from condensers to receivers should be based on liquid velocities of 100 fpm or less if only a single line is used. Recommended procedure of installing discharge lines and condenser to receiver lines, are shown

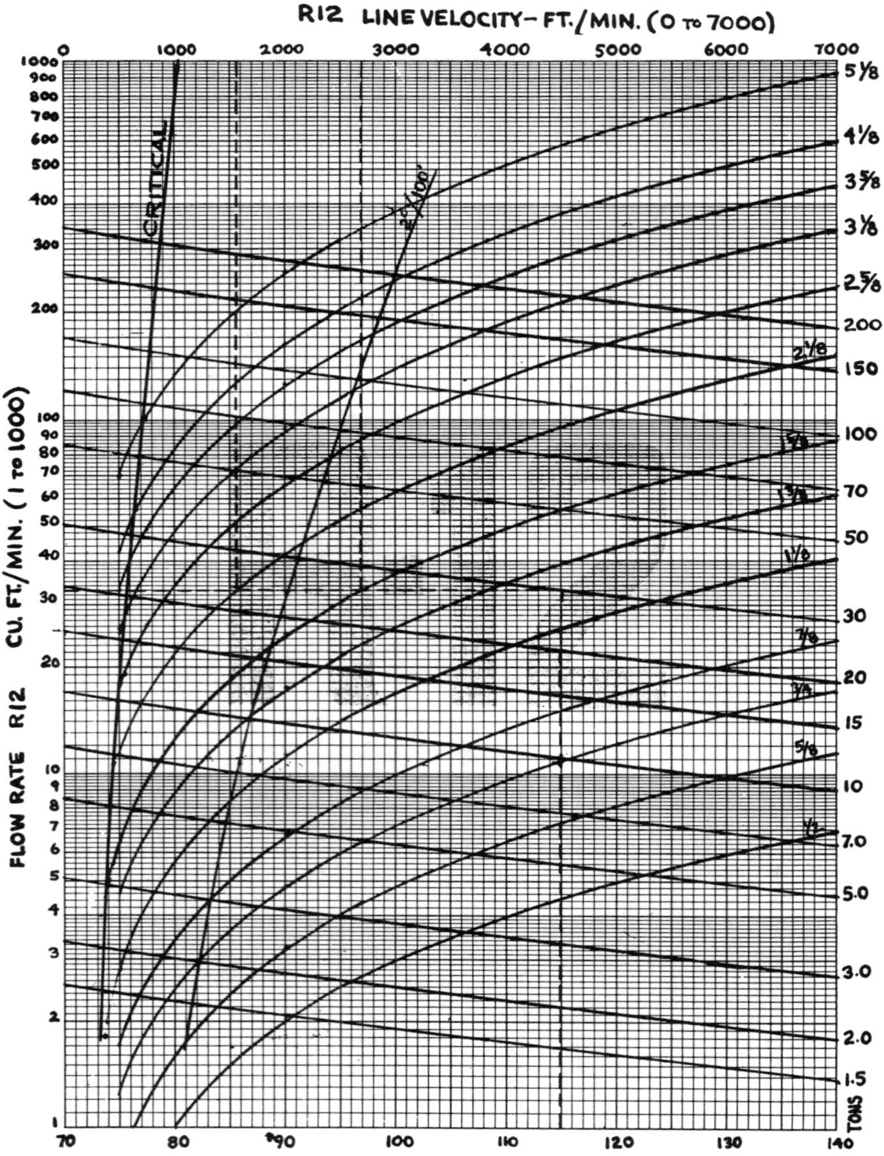
R12 LINE VELOCITY– FT./MIN. (0 TO 7000)
0
1000
2000
3000
4000
5000
6000
7000
FLOW RATE R12 CU. FT./MIN. (1 TO 1000)
CRITICAL
2°/100'
5⅛
4⅛
3⅝
3⅛
2⅝
2⅛
1⅝
1⅜
1⅛
⅞
¾
⅝
½
200
150
100
70
50
30
20
15
10
7.0
5.0
3.0
2.0
1.5
TONS
70
80
90
100
110
120
130
140

Fig. 50

in Fig. 51 in which case the liquid lines need not be larger than the one to the expansion valve or valves. Fig. 51 shows an equalizing gas line from the top of the receiver to the condenser inlet line.

For 100 tons capacity with 130° condensing, a 1⅜ tube will cause a friction pressure loss of 10 psi per 100 feet of line. This will be a satisfactory selection with the receiver over 15 feet above the expansion valves and the horizontal run not exceeding the vertical drop. With the receiver less than 10 feet above the expansion valves a 1⅝ line should be used. The gas equalizing line may be about half the inside diameter of the liquid line so that a ⅝ OD line would be satisfactory. The chart of Fig. 52 is used to select R-22 liquid lines and Fig. 53 is used to select R-12 liquid lines.

The R-12 liquid line at 100 tons and 130° condensing is shown as being between 1⅜ OD and 1⅝ OD for 10 psi friction loss per 100 feet, so the 1⅝ OD line should be used for R-12 if the horizontal run does not exceed the vertical drop.

No trouble results from using oversized liquid lines except the cost penalty.

Chilled Water Pump Selection

An accurate estimate of the length of the chilled water piping and the number and types of fittings required should be made and the equivalent length calculated. From this the friction loss in the piping can be calculated. The friction loss of the coils and the chiller should be

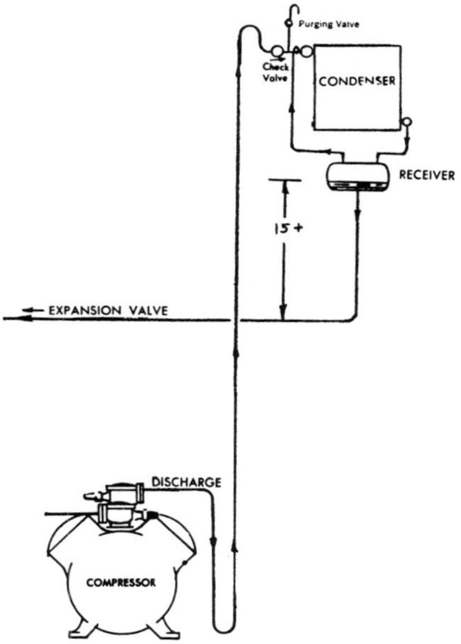

Fig. 51

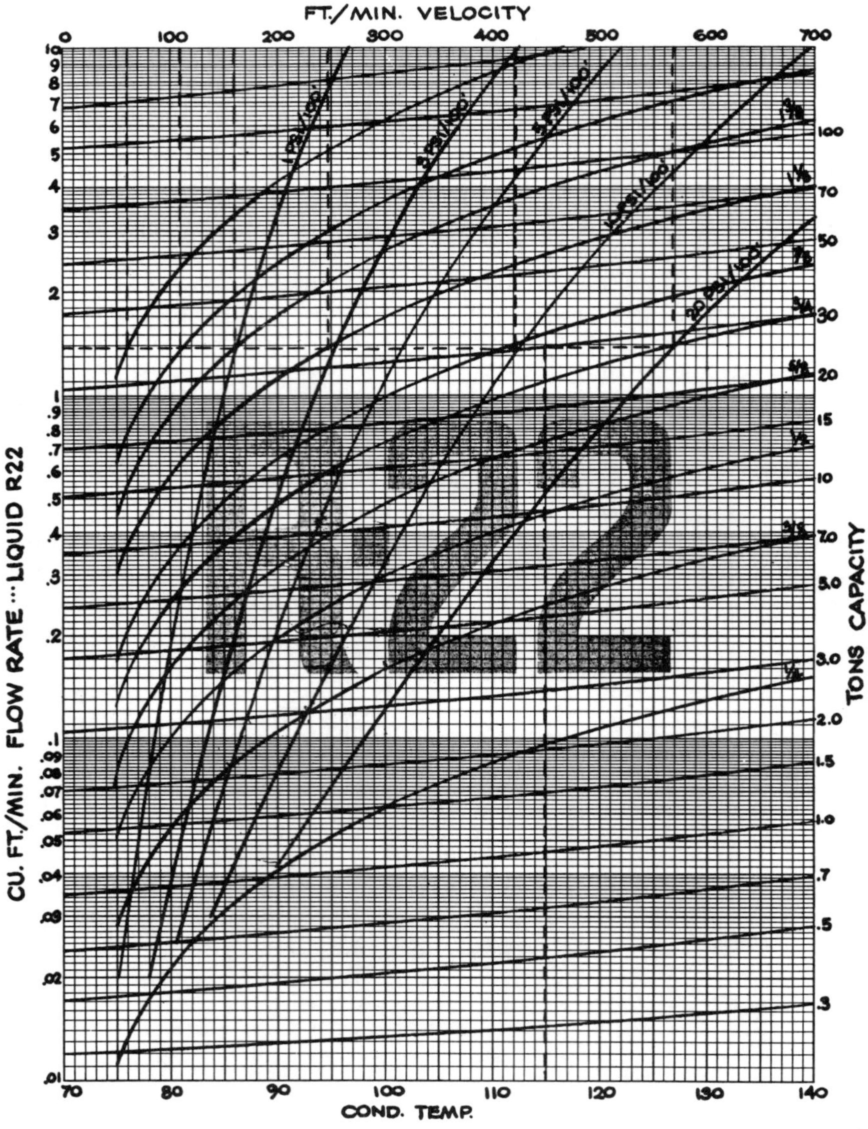
FT./MIN. VELOCITY
CU. FT./MIN. FLOW RATE ... LIQUID R22
TONS CAPACITY
COND. TEMP.
1 PSI/100'
2 PSI/100'
5 PSI/100'
10 PSI/100'
20 PSI/100'
R22

Fig. 52

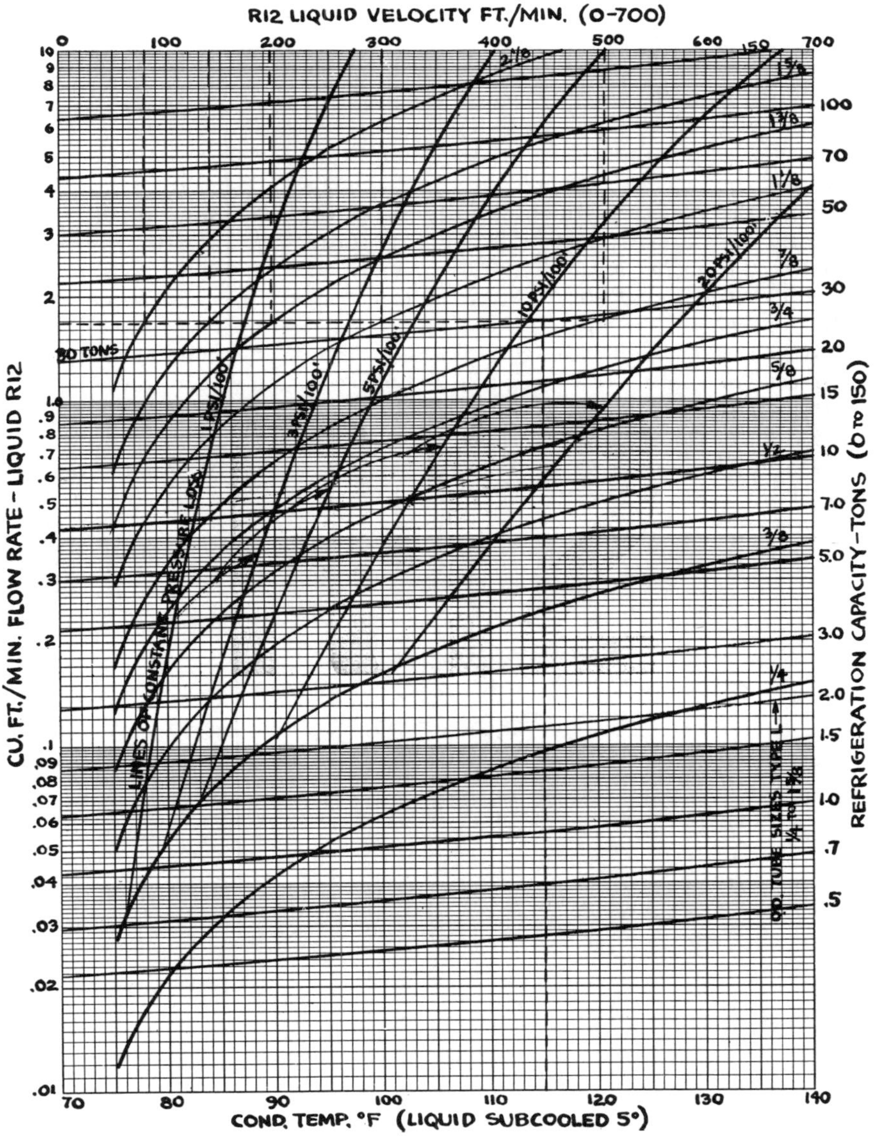

R12 LIQUID VELOCITY FT./MIN. (0-700)
0
100
200
300
400
500
600
700
CU. FT./MIN. FLOW RATE - LIQUID R12
REFRIGERATION CAPACITY-TONS (0 TO 150)
COND. TEMP. °F (LIQUID SUBCOOLED 5°)
70
80
90
100
110
120
130
140
30 TONS
LINES OF CONSTANT PRESSURE LOSS
1 PSI/100'
3 PSI/100'
5 PSI/100'
10 PSI/100'
20 PSI/100'
O.D. TUBE SIZES TYPE L
1/4 TO 2 1/8
2 1/8
1 5/8
1 3/8
1 1/8
7/8
3/4
5/8
1/2
3/8
1/4
150
100
70
50
30
20
15
10
7.0
5.0
3.0
2.0
1.5
1.0
.7
.5

Fig. 53

obtained from the chiller and coil manufacturers data and the total system pressure loss determined. With the rate of flow required in gpm and the total friction head in mil inches per foot the pump can be selected to match the requirements.

Cooling Tower Pump Selection

The selection of the cooling tower pump is essentially the same problem as selecting the chilled water pump except that a static head is added to the friction loss of the equivalent length of piping, the pressure loss through the condenser and, (if the cooling tower uses spray nozzles), the pressure required to provide the proper spray pattern must be added to the piping and other friction losses.

The static head in addition to the friction losses is the difference in altitude between the level of the nozzles in the cooling tower and the surface of the water in the storage tank. Fig. 54 shows the general piping arrangement for cooling tower piping including an automatic tower by-pass valve to prevent excess cooling by the tower on days where the ambient wet bulb temperature is low.

The automatic by-pass valve may be installed to bypass water from the condenser-to-tower line to the return from tower line. In this case the valve may be pressure operated responsive to discharge pressure.

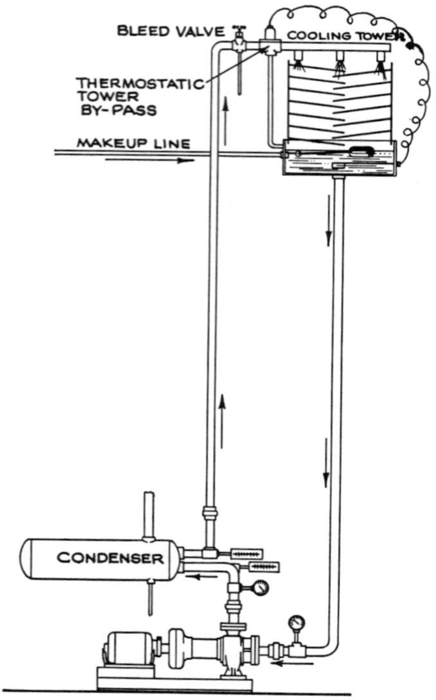

Fig. 54

If the valve is to be responsive to cooling tower tank temperature, the by-pass pipe must bypass the nozzles but not the tank.

Ventilation

Electric package liquid chillers reject heat to their surroundings. If proper ventilation is not provided, the machinery room temperature may rise to such a point that motor overload protectors may be affected causing unnecessary service interruptions. In extreme cases high machinery room temperature will contribute to motor burn-outs and other trouble associated with high ambient temperature.

For motors having 90% efficiency 10% of the electrical input is liberated as heat by the motor and the input per horsepower of motor output is the equivalent of 2830 Btu per hour.

A compressor operating at 38° evaporation and about 106° condensing has a mechanical efficiency of about 75%. This means that 25% of the compressor input goes to frictional heat. Of this heat, part is rejected to the machinery room and part is passed into the refrigerant and transferred to the condenser cooling water.

Assuming half of the frictional heat is rejected to the machinery room, the ventilation air required to remove the motor heat and half of the compressor frictional heat amounts to about 30 cfm per compressor horsepower if the ventilation air rises 20° in temperature as it passes through the machinery room. On this basis, for a machinery room housing a water cooled package liquid chiller of 150 horsepower, 4500 cfm will be required to maintain the machinery room at 20° above ambient if all machinery room heat is removed by ventilation air.

Wiring

Power wiring should be installed in accord with all local and national codes. Table 9 shows recommended minimum wire sizes for 220 or 440 volt power wiring to be supplied to units used for water chilling only. Where control panels include motor starters for chilled water pump or cooling tower pumps and fans, the wiring to the control panels should be sized to include total loads. Fuses should also be based on total loads.

The wire sizes recommended are based on runs of less than 100 feet between the power panel and the control panel of the unit and on 3 wires per conduit. For longer runs, an increase in wire sizes may be required.

A fused disconnect switch within sight of the unit must be provided, preferably, as close to the unit as possible.

Where fans and pumps are located out of sight of the unit, disconnect switches should be located within 50 feet and in sight of the fan and pump motors.

Hydronic Accessories

Items involved in hydronic cooling and heating systems include the heat exchangers, piping, pumps, means to mechanically refrigerate water

Table 9—Wire and Fuse Sizes

Motor Rating hp	Full Load Amps. 220V.	Full Load Amps. 440V.	Fuse Rating 220V.	Fuse Rating 440V.	Wire Size 220V.	Wire Size 440V.	Locked Rotor Amps. 220V.	Locked Rotor Amps. 440V.	Inc. Start Amps. 220V.	Inc. Start Amps. 440V.	115.V Control Trans. Rating
25	66	33	175	90	4	8	372	186	242	121	.5KVA
30	78	39	225	110	3	6	444	222	288	144	.5KVA
40	104	52	300	150	1	6	588	294	382	191	.5KVA
50	130	65	350	175	0	4	720	360	468	234	.5KVA
60	154	77	400	225	00	3	875	438	568	284	.5KVA
75	192	96	500	250	0000	2	1090	545	708	354	.75KVA
100	246	123	700	350	350MCM	00	1470	735	960	480	.75KVA

NOTE: Dual units use 2 motors and require two power circuits. Select fuse and wire size for each motor from above table. Control circuits are all 115V. and a separate fused disconnect switch is recommended for feed lines. Where the control circuit is to be taken off the power lines, the KVA rating of the 115V. secondary transformer is indicated in the last column of the above table.

during the cooling season and to heat it during the heating season.

In addition, a number of auxiliary devices are helpful in producing trouble-free, safe, and effective cooling and heating systems. These auxiliary devices include:

1. Expansion tanks
2. Air separators
3. Automatic fill valves
4. Pressure relief valves
5. Flow indicators
6. Miscellaneous fittings
7. Zone control valves

Expansion Tanks

An expansion tank is needed because the water in the hydronic system expands and contracts as its temperature rises and falls. Table 10 shows the density of water at temperatures of 40°F to 500°F expressed in pounds per gallon. During operation in a cooling cycle the limits of the chilled water temperature may be from 40° to 60° but during shut down periods the system water could conceivably rise to 80°.

A low temperature hydronic heating system operates at water temperatures up to 250°F. Conceivably, the temperature of the water could go down to 40° when out of operation.

For the cooling range of 40° to 80° the density of the water varies from 8.345 pounds per gallon to 8.314 pounds per gallon. The percent of change is .031/8.314 = .373 percent.

For the heating range of 40° to 250° the density varies from 8.345 pounds per gallon to 7.936 for a percentage variation of .409/7.936 = 5.15 percent. Thus the volume variation for the low temperature hydronic heating range is 5.15/.323 = 15.9 times as great as for cooling.

TABLE 10. Saturated Water Density Pounds Per Gallon

Temp. °*F*	*Density Lbs/Gal.*	*Temp.* °*F*	*Density Lbs/Gal.*
45	8.345	220	7.972
60	8.334	240	7.901
80	8.314	260	7.822
100	8.289	280	7.746
120	8.253	300	7.662
140	8.207	350	7.432
160	8.156	400	7.172
180	8.098	450	6.892
200	8.039	500	6.533

The expansion tank for a heating system is sometimes specified as 22 percent of the water volume in the system and has been shown as 1.5 percent to 3 percent of the volume of the cooling system. To determine the actual water volume change or the recommended expansion tank size it is necessary to estimate the volume of water in the piping, the liquid chiller, and the cooling coils for cooling. For heating the water capacity of the boiler is required also.

TABLE 11. Gallons per 100 Ft. Of Pipe

	Steel Pipe		*Copper Tube Type*		
Nominal Size	*Std. Sch. 40*	*Ext. Heavy Sch. 80*	*K*	*L*	*M*
¼	.541	.372	.379	.404	.431
⅜	.992	7.30	.660	.753	1.32
½	1.58	1.22	1.13	1.21	1.94
¾	2.77	2.25	2.27	2.50	2.68
1	4.49	3.74	4.05	4.42	4.54
1¼	7.77	6.66	6.34	6.55	6.81
1½	10.58	9.18	8.94	9.25	9.50
2	17.4	15.3	15.7	16.1	16.4
2½	24.9	22.0	24.2	24.7	25.4
3	38.4	34.3	34.5	35.4	36.2
3½	51.4	46.2	46.8	47.8	48.9
4	66.1	59.7	60.7	62.3	63.4
5	104.	94.5	94.0	97.1	98.1
6	150.	135.	135.	139.	142.
8	260.	237.	234.	243.	247.

To estimate the amount of water in the piping the length of each size of pipe in the system should be determined. Table 11 shows the gallons of water per 100 feet of various types and sizes of pipe and tubes.

To determine the quantity of water contained in the chiller itself calculations have been made on most evaporators. Based on the difference in volume between the inside of the casing and the volume of the evaporator tubes the smallest evaporator will hold about 4.8 gallons of water which is 63% of the casing volume. The largest evaporator holds about 58.2 gallons which is 50% of the casing volume. Based on the tonnage of the evaporators the water volume is approximately .3 gallons per ton of refrigeration capacity for the largest evaporators and varies to about .5 gallons per rated ton in the 60 ton range.

Since the water volume in the piping is apt to be high with respect to chiller water volume the percent error will not be great if the chiller volume is estimated as .5 gallons per ton of rated capacity.

With chilled water coils using 5/8" tubes and 6 rows deep, it is estimated that the water volume per ton of coil capacity will be in the range of .3 gallons per ton of refrigeration capacity. This can be widely variable depending upon the number of fins per inch, the spacing between tubes, water and air velocities, etc. Here again the volume is apt to be small in relation to the hydronic piping volume. Unless the coil volume is known a volume of .3 gallons water capacity per rated ton of coil capacity may be used for estimating.

Table 6 and Fig. 17 illustrates a 100 x 300 room using two 75-ton air conditioners connected to a penthouse installation of a 150-ton package liquid chiller. To determine the expansion tank requirement for such a system the water holding capacity can be estimated as follows:

Table 12

Pipe Run	*Pipe Size*	*Gal/100 ft.*	*Length ft.*	*Gallons*
A-B	5	104	60	62.4
B-C	4	66.1	166	109.7
C-D	4	66.1	15	9.9
EF	4	66.1	15	9.9
F-G	4	66.1	166	109.7
G-K	5	104	45	46.8
+K-A	5	104	25	26.0
+H-G	4	66.1	15	9.9
+B-J	4	66.1	15	9.9
150 Ton Chiller	at .50 per T			75.
2–75 Ton Coils	at .30 per T			45.
			TOTAL	514.2

The difference in volume for 514.2 gallons in a chilled water system with all water at 80° at one time and all water at 40° at another would be 514.2 gallons x .00323 = 1.66 gallons and the recommended expansion tank size would be 514.2 x .015 = 7.7 gallons or more.

For heating systems a water feed valve is often set to charge water into a piping system at a 12psig pressure. If the system is filled completely with water at an atmospheric pressure of 14.5 psia with the expansion tank containing only air at that pressure and holding 7.7 gallons the air would be compressed if its pressure was raised to 12 psig or 14.5 + 12 = 26.5 psia.

If no temperature change occurs in the tank the product of the initial absolute pressure and volume of the air is equal to the final pressure volume product or $P^1V^1 = P^2V^2$ where P^1 is the initial absolute pressure, P^2 is the final absolute pressure, V^1 is the full tank volume, and V^2 is the air volume after raising the pressure to 12 psig.

$$14.5 \times 7.7 = (14.7 +) V^2$$

$$V^2 = \frac{14.5 \times 7.7}{26.5} = 4.2 \text{ gallons}$$

If the water fill line was turned off and the water temperature was reduced from 80° to 40°, 1.66 gallons would be withdrawn from the tank and the air volume would then be 4.2 + 1.66 = 5.86 gallons and the pressure would be

$$P^3 = \frac{14.5 \times 7.7}{5.86} = 19 \text{ psia or } 19 - 14.5 = 4.5 \text{ psig}$$

If the water charge line was opened and recharged with water until the pressure is up to 12 psi with 40° water in the system the air volume would return to 4.2 gallons. If the chiller was shut down and the water warmed up to 80° again and the volume increased 1.66 gallons the volume of the air in the tank would be reduced to 4.2 — 1.66 = 2.54 gallons and the pressure would then be $\frac{14.5 \times 7.7}{2.54} = 44$ psia or = 44 — 14.5 = 29.5 psig.

A pressure relief valve on the chiller with a 30 psi setting would not open at this condition so that as the system water varied from 40° to 80° the expansion tank pressure would vary from 12 psi to 29.5 psi and no water would be added or discharged.

Expansion Tank Location

Fig. 55 is a sketch showing water in a pipe loop containing a chiller, a pump, and a cooling coil.

The expansion tank is shown connected at the pump outlet. With the piping all on one plane and the pump "off" the whole system will be at the expansion tank pressure. When the pump starts the discharge of the pump will still be at the same expansion tank pressure. However, the pump was to develop a 40 ft. head and since the discharge pressure at 40° is 12 psi the suction pressure at the pump would be 12 psi less the head supplied by the pump or $12 - \frac{40}{2.31} = 12 - 17.3 = -5.3$ psi or 10.8 inches vacuum. (2.31 ft. of head = 1 psi)

This is not an acceptable arrangement and any suction side leaks would allow air to be drawn into the system.

Fig. 56 with the expansion tank connected to the suction side of the pump would always maintain a 12 psi pressure (with 40° water) at the suction side of the pump and with a 40 ft. pump head when running, the discharge pressure of the pump would be 12 psi + $\frac{40}{2.31}$ = 12 + 17.3 = 29.3 psi

With the water at 80° and using an expansion tank volume of 1.5 percent of the water volume the expansion tank and pump suction pressure would increase from 12 to 29.5 psi and the pump discharge pressure would increase to 46.8 psi. This system is entirely satisfactory from the standpoint of pressures throughout the system. The water leaving the chiller is at its lowest temperature and lowest pressure.

For air separation from water the ideal location is at the point of lowest pressure and highest temperature so that the ideal arrangement is as shown in Fig. 57.

In this case the suction pressure will be the same at the pump as in the previous case and the pump discharge pressures will be the same. However, the return water temperature to the pump and chiller will be higher (during operation) than any other point so that this provides an ideal air separator location. Actually the temperature change is so small in the chilled water system that Figs. 56 and 57 are equally satisfactory.

In high rise systems the point of lowest pressure and high temperature may be at the top of the return piping. In such cases it may be advantageous to have the air separator and expansion tank at the top of the return piping and locate the pump to discharge into the liquid chiller.

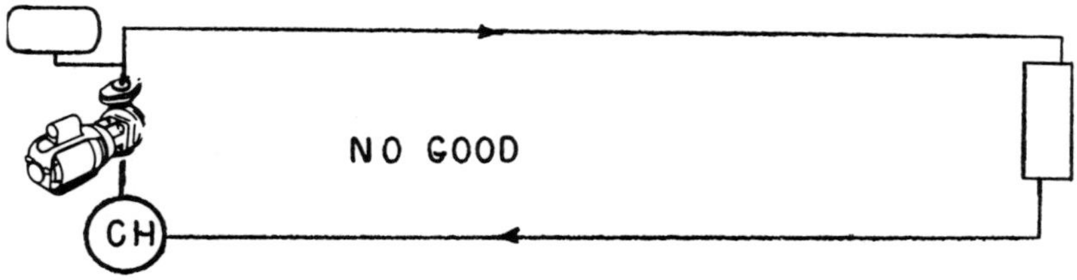

Fig. 55

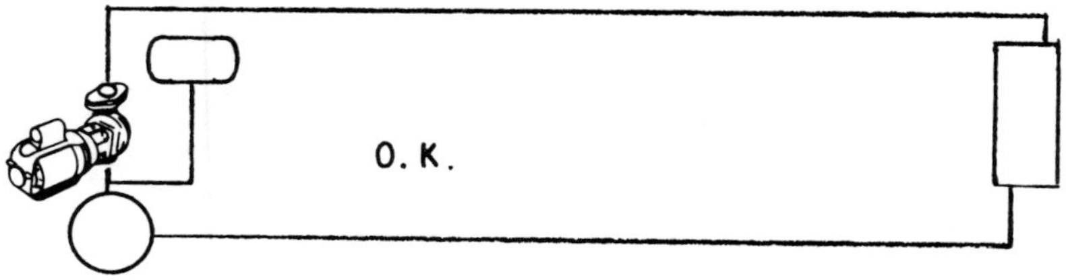

Fig. 56

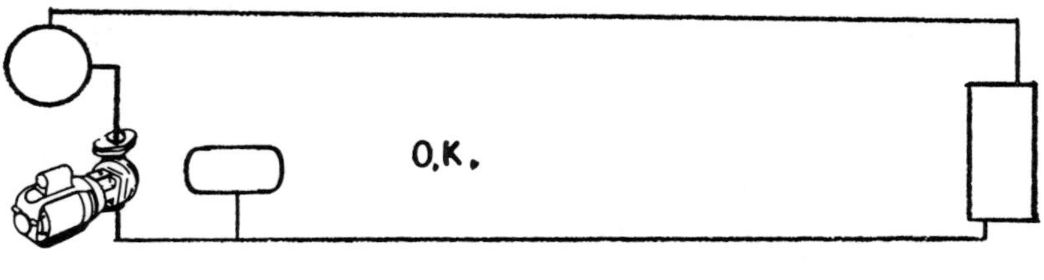

Fig. 57

In this case a minimum pressure of 4 psi is recommended at the top or lowest pressure point. The pump suction pressure will be higher than this by the difference between the static pressure increase due to static head and the friction loss in the return pipe.

Air Separators

Fig. 58 shows an air separator in which the incoming water, containing entrained bubbles, shows down in the shell of the separator allowing the bubbles to rise. The outgoing water is taken from near the bottom of the casing to the pump suction pipe. The air rises around the dip tube and is carried to the compression tank by a pipe not less than ¾ inch nominal diameter.

Fig. 59 shows an air separator in which the incoming water enters tangentially to provide a swirling action. The swirling water leaves tangentially from the lower water connection to the suction side of the circulating pump. The air being lighter goes to the center of the swirling water and enters the perforated tube which connects to the expansion tank.

Fig. 60 shows an expansion tank fitting. This fitting provides 3 pas-

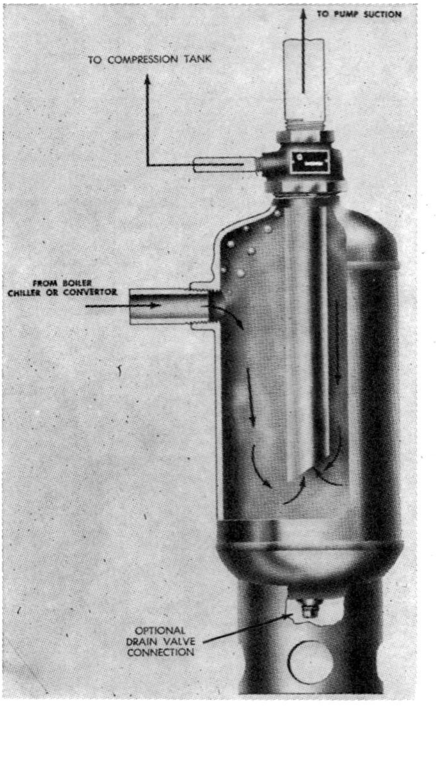

Fig. 58

Fig. 59

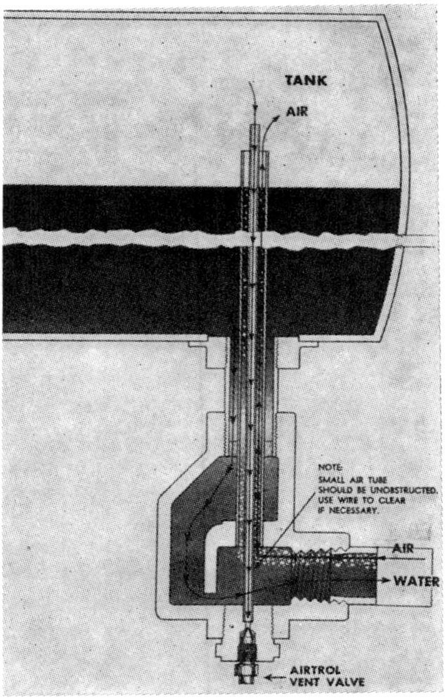

Fig. 60

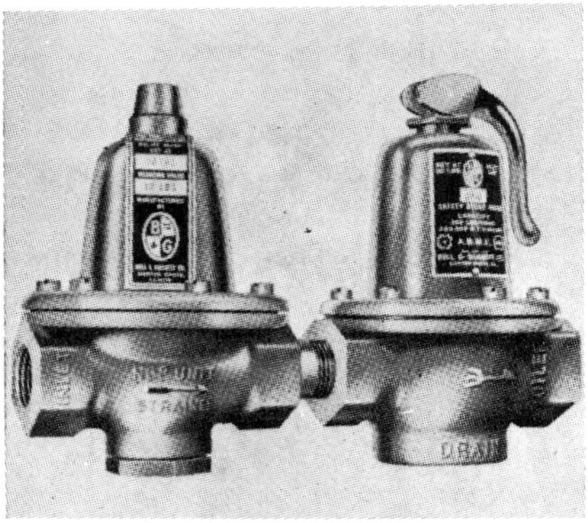

Fig. 62

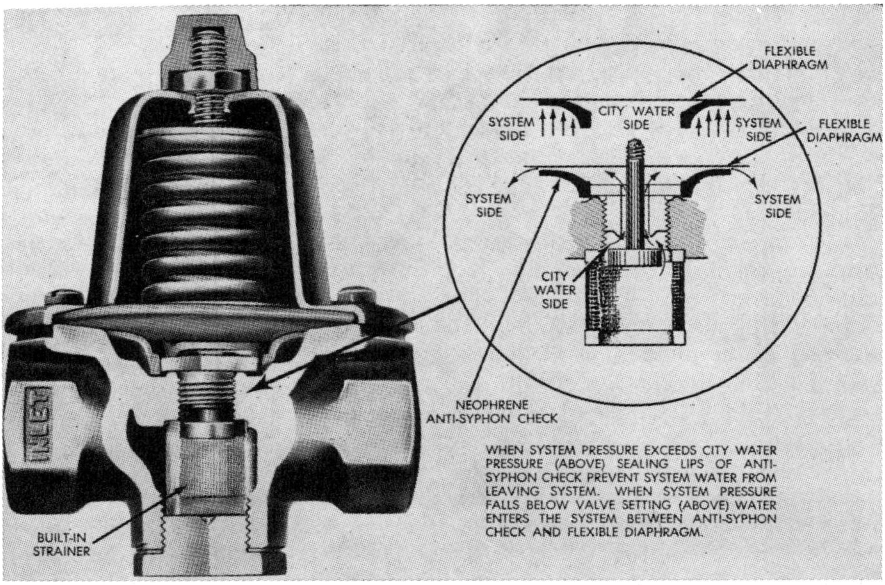

Fig. 61

sages into the expansion tank. The outer passage has free access into the bottom of the tank for water as it expands in the system.

The air is trapped out of the water by the submerged baffle and passed up the outer of two concentric tubes into the air section of the tank. The small tube in the center is connected to a vent valve at the bottom of the fitting. This valve may be manually opened to vent out air to reduce the pressure in the expansion tank.

Fig. 61 shows a water pressure reducing valve. The inlet is supplied by water from a fresh water supply, usually city water which must have a pressure above the charging pressure of the hydronic system. The valve may be adjusted to feed water at any pressure it is set for below the inlet water pressure. The valve opens whenever the force of the water on the lower side of the diaphragm is less than the spring force on top of the diaphragm. If the force on the bottom is greater than the spring force the valve remains closed.

This valve has a built in check valve so that the water flow cannot reverse even though the inlet water pressure drops below the system pressure and the system pressure is low enough to be overcome by the spring force. This check valve prevents the city lines from being contaminated with system water and protects the hydronic system against loss of water if the pressure in the supply line drops.

Fig. 62 shows a combination pressure reducing valve and a pressure relief valve. Each valve is adjustable. Cooling systems can operate without expansion tanks. If the combination pressure reducing and pressure relief valve is used at a high point in the system it could be adjusted to feed water into the system at any desired pressure such as 12 psi. Also the pressure relief valve can be set to discharge water and air at a safe pressure such as 40 psi.

With such a system, if the system pressure dropped to less than 12 psi as the chiller operated, the pressure reducing valve will open to supply additional water and maintain a 12 psia pressure as water is fed in to compensate for the contraction of system water due to temperature reduction with the chiller operating. With the chiller off and the system warming up the system water expands. With no expansion tank and air cushion the pressure will rise to 40 psi with a slight change in water volume and discharge a volume of water essentially equal to the water expansion. In effect such a system operates like a thermal hydraulic pump taking in water at 12 psi and discharging water at 40 psi as the water contraction and expansion takes place.

Saves Water

Such a system would not use an excessive quantity of water in a cooling system where the average system water temperature varies only a few degrees. In a heating system the temperature changes are normally greater and the quantity of water used would be many times greater.

An important advantage of maintaining a "closed" system is that rusting due to oxygen in the water is not a continuous process due to using up the oxygen in the water initially and not allowing more oxygen to enter. With the use of pressure reducing and pressure relief valves without an expansion tank, the system would take in fresh water with its oxygen each cycle so as to make the rusting of the pipes a continual process. Therefore such systems are not recommended.

Fig. 63 shows a thermoflo indicator which is a water flow indicator with an adjustment to regulate flow. With systems having more than one riser, these devices may be installed in each riser. They may be adjusted to provide the design flow rate in each riser and give a visual indication of the flow. In addition to being of considerable assistance in the original balancing of the hydronic system, they serve as aids in case of service problems.

The proper selection and installation of auxiliaries along with the correctly sized piping, pumps, chillers, chilled water coils, blowers, and controls, are all necessary to provide trouble free and effective hydronic cooling systems.

Cooling and Heating Service

Hydronic cooling or heating involves removing heat from or adding heat to a space by cold or hot water. In cooling especially, the heat is usually transferred from air to chilled water which carries it to the liquid chiller which in turn transfers the heat to refrigerant. The refrigerant transfers the heat to outdoor air or cooling water. In heating, the heat may be partially transmitted to air by means of heat exchangers and partially to objects by radiant heat transmission. This is especially true of radiant heating panels where no forced air circulation or convection air circulation cabinet is used.

The rate of heat transfer from air to water or vice versa in a heat exchanger is of primary importance in determining fauls in hydronic cooling or heating systems. In general, the factors affecting heat transfer between the two fluids are as follows:

(a) Area of the heat exchange surface.

(b) The temperature difference between the two fluids involved.

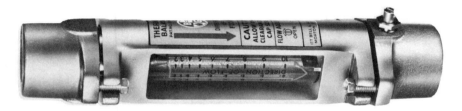

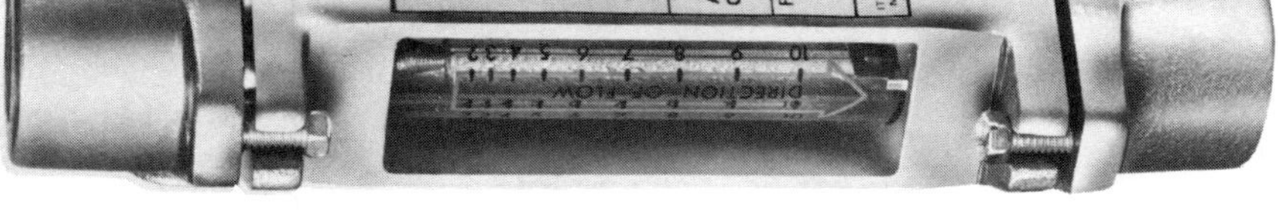

Fig. 63

(c) The velocity of each fluid.
(d) The characteristics of the fluids.
(e) The flow patterns. Streamline or laminar flow over a heat exchange surface exposes only the surface of the stream to the heat exchange surface. A turbulent flow mixes the fluid in a stream so that all of the fluid contacts the heat exchange surface.
(f) A "wet" surface produces a higher heat transfer rate than a dry surface from air passing over the surface.
(g) To a lesser extent, the heat conductivity of the metals normally used in heat exchangers and the thickness of the metals also affects heat transfer. The conductivity of all the metals used in general is so high relative to heat transfer from a fluid to a surface, that the temperature difference through the metal is very small, compared to the temperature difference between the two fluids. Consequently doubling or tripling the heat conductivity through the metal has little effect on the total temperature difference.

In determining the cause for a relatively low heat transfer rate, it is helpful to be able to measure the flow rate of both the water and the air in a heat exchanger. The total heat quantity going into a heat exchanger is equal to the total heat quantity out for steady flow conditions.

The specific heat for water is 1.0 Btu per pound of water per °F of temperature change, and a gallon of water weighs about 8-1/3 pounds so the heat transfer for water at any flow rate and temperature difference is:

Btu per hr = gpm x 8-1/3 lb. per G. x 60 min. per hr x °td.

When heating air or when cooling dry air the heat transfer rate is:

(F1) Btu per hr = gpm x 500 x °td.

Btu per hr. = cfm x D x 60 x °td x C.

where: cfm is the air flow rate in cubic feet per minute
D is the air density in pounds per cubic foot. At standard conditions this is .075 lb per cu ft.
60 is the number of minutes per hour
°td. is the difference in air temperature into and out of the heat exchanger
C. is the specific heat of the air which at constant pressure is .24 Btu per pound of air per °F of temperature difference.

For standard conditions, then:

(F2) Btu per hr = cfm x 1.08 x °td.

At other than standard conditions of pressure, temperature and humidity the density of the air in terms of pounds per cubic foot changes and so the above formula is only approximate.

Whenever air is being cooled by a heat exchanger whose surface temperature is below the dew point of the air passing over it some moisture will condense out of the air. In order to condense moisture its latent heat must be removed. Consequently in a chilled water air cooling coil part of the heat added to the water in the coil is latent heat from the water vapor condensed. Consequently only the balance of the heat added to the water is removed as sensible heat from the air.

The heat transfer rate in cooling air is stated as:

Btu per hr = cfm x D x $(h^i - h^o)$ x 60

where h^i is the enthalphy of the air into the heat exchanger in Btu per pound of air and h^o is the enthalpy of the leaving air.

(F3) Btu per hr = cfm x 4.5 x $(h^i - h^o)$ (for standard air)

The enthalpy of air may be determined from a psychrometric chart from two independent properties or from its wet bulb temperature.

Wet bulb temperature can be read from an ordinary thermometer or thermocouple which has its temperature sensing portion covered with a damp porous material such as a damp wick. This temperature may be difficult to obtain accurately and an error of one degree may represent an appreciable error in enthalpy.

Dry bulb temperature, relative humidity and dew point are also satisfactory as independent air properties to be used in determining enthalpy.

Water Flow Rate Measurement Water Meter — Time Readings

Various methods of measuring water flow rates are available but not necessarily convenient. If a water meter is located in the water line it can read, with varying degrees of accuracy, the total volume flow in a given period of time in which case the average flow rate in terms of gpm is the total flow in gallons divided by the number of minutes used to flow that number of gallons.

Water Flow Meters

Flow meters often consist of a restrictive orifice in a section of the pipe and a gauge to indicate the drop in pressure through the orifice. The pressure drop varies as some power of the flow rate and the meter may be calibrated in terms of gallons per minute.

Fig. 64 shows a type of flow meter which may be installed in parallel branch water circuits. It makes use of a spring held disc in the water flow so that the position of the disc is dependant on the flow rate through the tube. These devices also contain variable restrictors which may be adjusted to obtain the design flow rate in each of more than one parallel circuits.

By their very nature, a flow meter is not the type of instrument which can readily be applied to an existing system in the manner in which an electrical voltmeter may be used. Therefore, checking the flow rate through an individual fan coil unit of a system having several coils becomes a problem due to the piping involved.

Air Flow Measurement Anemometers

Air flow may be measured by an anemometer, Fig 65 or velometer Fig. 66. An anemometer may consist of a fan type blade driven by the air passing over it. The propellor type wheel works through gears to indicate the number of lineal feet of air that passes by the wheel. If the number of lineal feet of air passing by the wheel is divided by the time in minutes required for the passage the velocity is found in feet per minute. It is usually necessary to refer to a calibration curve to correct for friction at various velocities. The rate of flow is then determined from Q = AV where Q is the rate of flow in cfm, A is the cross sectional area of the air stream in square feet, and V is the average velocity in feet per minute (fpm). The air velocity may not be uniform in all parts of the air stream. An average is obtained by moving the anemometer around slowly over all sections of the airstream to obtain an average velocity.

Fig. 64

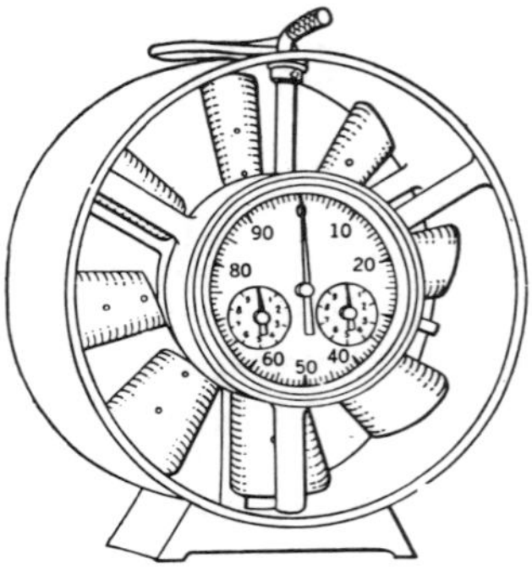

Fig. 65 Anemometer.

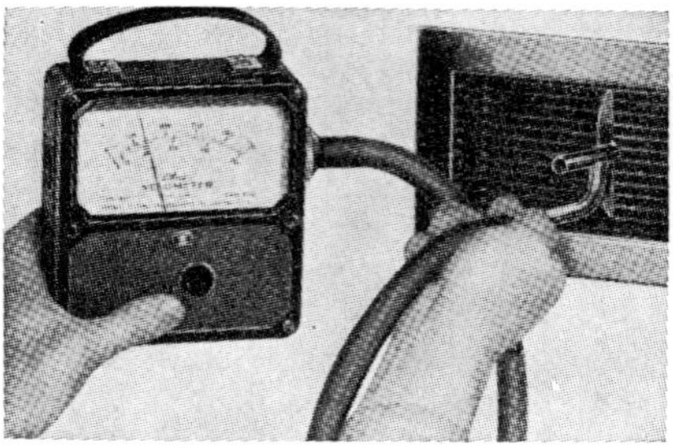

Fig. 66 Velometer.

Velometers

Several types of velometers are calibrated to give direct and essentially instantaneous readings of air velocity by measuring the velocity pressure of the air. The velocity pressure is the difference in total pressure, as obtained from a measuring tube pointing directly into the air stream, and the static pressure which is the pressure exerted on the walls of the ducts. The air velocity pressure is measured in inches of water and by various devices may be read in hundredths or thousandths of an inch of water.

The velocity of the air is $400 \times \sqrt{"H^2O\ (V.P.)}$

One inch of water gives a velocity $V = 4000\sqrt{1.0} = 4000$ fpm

One quarter inch velocity pressure indicates a velocity of $V = 4000\sqrt{.25} = 4000 \times .5 = 2000$ fpm. A velocity of 1000 fpm gives a velocity pressure of .0625 inches of water.

One type of direct reading velometer has a tapered air passage containing a movable vane. At no flow a hair spring returns the indicating hand to zero and the vane in effect closes off the air passage. As the air velocity increases some air is admitted through a calibrated orifice and blows through the tapered passage in the meter which causes the vane to move the indicating hand. This type of meter may have more than one orifice size to admit air to the tapered passage. This permits more than one velocity scale to be obtained with the same meter such as 0 to 500 fpm and 0 to 2500 fpm.

Hot Wire Anemometer

Another type of air velocity measuring device consists of two thermocouple junctions which are heated electrically. One is in the airstream while the other is in still air. The junction in the air stream loses heat at a rate depending on the air velocity by it. The resultant temperature difference between the two junctions is calibrated in terms of feet per minute air velocity. The scale can be changed readily and may be calibrated to read velocities as low as 10 fpm. A flexible lead has a probe which is held at any point in the airstream. The outlet area of a duct may be laid off in several sections and the velocity measured at the center of each section and recorded separately. The average velocity is then calculated to use with the outlet area to determine Q from Q = AV.

The flow of water through a heat exchanger can be calculated from water in and out temperatures, air flow and wet bulb temperatures, etc., by equating the heat in and heat out as follows:

Btu per hr = $\text{gpm} \times 500 \times {}^\circ\text{td} = \text{cfm} \times 4.5 \times (h^i - h^o)$ and

$$\text{gpm} = \frac{\text{cfm} \times 4.5 \times (h^i - h^o)}{500 \times (t^o - t^i)}$$

$$\text{(F4)}\quad \text{gpm} = \frac{\text{cfm}\ (h^i - h^o)}{111\ (t^o - t^i)}$$

where cfm = the measured air flow through the coil
h^i = enthalpy of entering air
h^o = enthalpy of leaving air
t^o = water temperature out
t^i = water temperature in

h^i and h^o are obtained from the psychrometric chart from inlet and outlet wet bulb temperatures or other determining factors.

Service Complaints

The majority of service complaints associated with hydronic cooling systems will fall under one of the following classifications:

1. Does not cool
2. Insufficient cooling
3. Air too damp
4. Temperature varies too much
5. Water dripping on floor
6. System noisy

The variation between hydronics systems is so great that only general analysis and suggested remedies for each of the above complaints are attempted here. In a general analysis it can be stated that:

(a) If the heat exchanger is of the required size
(b) It flows water at the required rate
(c) At the required water temperature
(d) And the required air flow over the coil

It will do a proper job of air cooling. Checking each of these factors is the basis of hydronic cooling service. In the following complaints, possible causes, and possible remedies are suggested.

Does Not Cool

1-a Chiller not operating. Start chiller if it has not been shut down for the season. If it has undergone a seasonal shut down, consult chiller service manual for start-up procedure.

1-b Chiller pump not operating. This pump must operate to keep compressor operating. Check overload contacts and controls interlocked with compressor.

1-c Zone control valve not positioned for cooling. If system uses hot or cold water in the same heat exchanger, controls will have to be checked to see if controls allow cooling zone control valve to admit chilled water to cooling coil. This problem is most apt to occur with seasonal weather changes.

1-d No air flow over coil. Check to see that fan blower is operating and any dampers involved are positioned for cooling.

Insufficient Cooling

2-a-1 Chilled water temperature not low enough. Check outlet temperature at chiller. If too high at chiller outlet set thermostat on package chiller for lower temperature. If control responds to return water temperature it should normally not be set for lower than 50°F return. The supply should be between about 40° to 46° for maximum cooling.

2-a-2 If the water temperature is satisfactory at the chiller but is too high entering the coil, check to see if warm water can be mixing with cold water. Also check temperature of supply pipe at various points along the run. The system should provide the same water temperature at each takeoff point along the line. If it is installed as a one pipe system with supply and returns tieing into the same pipe the chiller may be set to provide water at 40°F in an attempt to provide satisfactory cooling in downstream units.

2-b-1 Water flow rate is too low through coil. Check for balancing valve opening. Adjust if necessary to get proper flow rate. Flow rate can be checked as per formula. (F4) The flow should be not less than 2 gpm per ton of cooling required from the coil.

2-b-2 If the chilled water flow rate is too low in the entire system the chilled water pump should be checked. Checking the electrical input to the pump will indicate if the pump is underloaded or normally loaded. Underloading is apt to be the result of high pressure difference across the pump which can also be noted from inlet and outlet gauges. Low system flow will result in greater than normal temperature drop through the chiller when it is fully loaded. If all indications point to low flow for the pump used, the pump should be examined for impeller diameter and condition.

2-c Heavy latent heat load. The cooling load in cooling air from 80° to 75° with very humid air is about the same as cooling dry air from 80° to 60° and the actual temperature reduction of only 5° may make the apparent cooling appear to be low. Check for reason for high humidity air. Check ventilation air supply and condition.

2-d Insufficient air flow. Check air flow through the cooling coil. An air flow rate of 300 to 400 cfm per ton of cooling required should be ample.

3-a Water Temperature too High. Set water temperature lower. Open bypass dampers and close down coil face dampers. Activate reheat coils. In any building moisture is added by plants, people, gas burning equipment, cooking, and fresh air. This is usually removed by air exhaust and by condensation on the cooling coil. Lowering the water temperature to provide faster dehumidification also removes more heat and if the temperature becomes too low adding only sensible heat in a reheat coil corrects the condition.

3-b Air delivery rate too great for load. Lower air delivery rate, apply reheat or open coil bypass. If the air delivery rate through a cooling coil is too great for the load the room will approach the relative humidity of the air from the cooling coil unless face and bypass dampers are used or reheat is supplied.

Temperature Varies Too Much

4-a-1 Incorrect location or adjustment of room thermostat. Check thermostat location. If the thermostat is so located that it can receive direct or reflected sunlight at any time during the day, it will call for additional cooling which may cause a compensating drop in room temperatures as long as the thermostat receives the extra radiant heat.

4-a-2 The cooling thermostat differential may be too great. Usually a 2° temperature differential in the thermostat may be commercial and an anticipating means must be added to reduce the operating differential to ½° or less.

4-b Overcapacity Fan Coil Units. Lower air velocity over coil. For highly overcapacity individual fan coil too much cooling may be done during one "on" cycle of the fan. Reducing the fan speed, "dampering off" the air flow, providing more ventilation air, or decreasing thermostat differential may be helpful.

4-c Too much Variation in Water Temperature. This can be due to too much differential in the chiller control or failure of capacity control equipment to unload.

Water Dripping on Floor

5-a This is characteristic of fan coil units when control is used to stop and start the fan. With the fan off, the drain pan chills to below the dew point and sweats. Adding insulation to outside of fan coil unit may be practical or to inside of coil cabinet if it sweats.

5-b Drain pan overflows. This may be due to traps in the drain line or plugged drain lines. Determine cause and apply remedy which may suggest itself.

5-c Leaks in drain pan or piping. Locate and repair leaks. Also examine chilled water connections, piping, and purge valves for location of leaks. Repair as required.

5-d Inadequate or no isulation on chilled water piping. Insulate supply and return chilled water lines with adequate insulation.

System Noisy

6-a Air in system. If air is circulating with the water a gurgling noise may be apparent as the bubbles flow through restricted openings or the pump. It may be necessary to purge excess air out of the expansion tank. Also check for air separating equipment to be sure that it is properly installed.

6-b A "coffee grinder" sound in the pump may be due to worn bearings in the pump or motor. If so the bearings will have to be replaced.

6-c Pump vibration. Pump vibration may be due to faulty impellers which have erroded or corroded to cause imbalance. Also check couplings and be sure motor and pump are properly aligned (where applicable). Also check motor rotor for balance. Repair or replace as necessary.

6-d A squeal or shriek in a piping system may be due to excessive flow through a restricted opening in the piping. Check system for restrictions. Also check pump motor for current input as an indication of possible overloading or underloading. Check pressure difference across pump.

Also check motor and pump bearings for lubrication.

To a large extent trouble indications must be analyzed as they occur and the appropriate remedy will suggest itself to an experienced technician. The above suggestions may be helpful in a very general way only.

For servicing the package liquid chiller itself, consult the manufacturers service manual for electric package chillers, or for natural gas engine driven package chillers.

INDEX